SUPPLY CHAIN MANAGEMENT AND LOGISTICS

INNOVATIVE STRATEGIES AND PRACTICAL SOLUTIONS

SUPPLY CHAIN MANAGEMENT AND LOGISTICS

INNOVATIVE STRATEGIES AND PRACTICAL SOLUTIONS

EDITED BY

ZHE LIANG

WANPRACHA ART CHAOVALITWONGSE

LEYUAN SHI

CRC Press

Taylor & Francis Group

Boca Raton London New York

CRC Press is an imprint of the
Taylor & Francis Group, an **informa** business

CRC Press
Taylor & Francis Group
6000 Broken Sound Parkway NW, Suite 300
Boca Raton, FL 33487-2742

First issued in paperback 2017

© 2016 by Taylor & Francis Group, LLC
CRC Press is an imprint of Taylor & Francis Group, an Informa business

No claim to original U.S. Government works

ISBN-13: 978-1-4665-7787-9 (hbk)
ISBN-13: 978-1-138-89325-2 (pbk)

Contents

Section I Supply Chain Strategy and Coordination

Section II Supply Chain Network Optimization

Section III Inventory Management in the Supply Chain

List of Figures

List of Tables

Preface

As the world moves toward more competitive and open markets in the twenty-first century, effective supply chain management is of critical importance to the success of an enterprise. Despite a large amount of research conducted in the past decades on the supply chain management topic, many researchers and practitioners are still devoting considerable efforts to the new emerging problems. This is not only due to theoretical and computational challenges, but also the business environments and configurations from various industries, which are continuously changing and becoming more restrictive and demanding. As a result, numerous new problems are arising in the field of supply chain management.

In response to the need for educational and research resources that practically apply to a collaborative and integrative environment in today's market, this book contains contributions from leading experts in supply chain management throughout the world. It is intended as a collection of innovative strategies and practical solutions that address problems encountered by enterprises in the management of supply chain and logistics. As supply chain management is a far-reaching area, it is not possible to cover all aspects and applications of the field. Rather than concentrating on just methodology or techniques (such as optimization or simulation) or specific application areas (such as inventory or transportation), the book is designed to present readers with a collection of topics that bridge the gap between the academic arena and industrial practice. Yet the book still provides an in-depth discussion of both general techniques and specific approaches to a broad range of important, inspiring, and unsolved questions in the field.

This book is designed to be of value to researchers, practitioners, and professionals in academic institutions and industry who need a wide-spectrum resource for many different aspects involved in supply chain management from technical methodologies to management implications. Graduate (and advanced undergraduate) students and researchers will also find this book a rich resource for the design, analysis, and implementation of supply chain management problems arising in a wide range of industries.

The book is organized based on four major research themes in supply chain management: (1) supply chain strategy and coordination, (2) supply chain network optimization, (3) inventory management in the supply chain, and (4) financial decisions in the supply chain. The sequence of these themes helps in transitions from an enterprise-wide framework to network design to operational management to financial aspects of supply chain. Each individual theme also addresses the answer to a challenging question as to how to go about applying quantitative tools to real-life operations, resulting in practical solutions.

The first section includes two chapters focused on supply chain strategy and coordination. Chapter 1, by Miller, lays down the platform of this book by providing an overview of the concept of hierarchical supply chain planning frameworks at the strategic level down through operations. Although the chapter does not review the details of key operations and decision support tools in the frameworks, it provides a key foundational tool to organize and manage supply chain planning and operations activities. Chapter 2, by Pei et al., provides an overview of coordination between supply chain partners in discrete manufacturing enterprises. The chapter provides an overview of challenging scheduling problems that often arise in supply chain coordination and briefly discusses research methodologies that are applied to these problems.

The second section includes three chapters with emphasis on optimization methodologies in supply chain networks. Chapter 3, by Lei et al., provides an overview of optimization models and solution methodologies for integrated operations planning problems in supply chain networks. These ideas can be applied to integrated real-life problems that involve production, inventory, distribution, and routing. Chapter 4, by Nurre et al., presents a review of the nested partitions method for solving large-scale optimization problems. The applications of the method are demonstrated on two supply chain network optimization problems. Specifically, the intermodal hub location problem, which is a facility location problem in supply chain networks, and the multilevel capacitated lot-sizing problem with backlogging, which is a complex production planning problem, are the two case problems in the chapter. Chapter 5, by Chen and Shi, considers a novel stochastic optimization model of the location problem to place critical components in a supply chain network to increase the resiliency of the network, that is, to aid in the recovery of the supply chain network after an extreme event. The chapter discusses a case study to determine the placement of permanent generators at the retail locations of shops, which distributes both convenience items and fuel in Upstate New York and Vermont.

The third section includes two chapters that are centered on inventory decisions in the supply chain. Chapter 6, by Liang et al., proposes a novel optimization model based on a schedule-based formulation for the cyclic inventory routing problem. The chapter presents a column generation method as a solution methodology to solve this problem efficiently. Chapter 7, by Chaovalitwongse et al., illustrates a production planning case study in a production process of rolled tissues. In the case study, customer demand could not be fulfilled because of an insufficient inventory level. The chapter addresses the problem by introducing a new inventory policy and a production planning method to determine new production orders.

The fourth section includes two chapters that are focused on financial aspects in the supply chain. Chapter 8, by Yang et al., discusses the importance of competitive learning when firms make decisions on initial investments. For example, when firms neglect the due diligence on conducting demand-forecast studies, overcapacity will be inevitable. The last chapter, by Qin et al., studies a revenue management optimization for a multiperiod multiclass rail passenger revenue management problem. The chapter presents a complex optimization model and proposes a new efficient solution methodology. This work has an impact on logistics network and transportation problems.

During the process of completing this volume, we spent a few years interacting with the authors and anonymous reviewers. We appreciate their time, effort, and dedication toward the successful completion of this volume and cannot thank them enough. The experience of putting together this volume has been rewarding. We truly hope that readers will find the volume to be as stimulating and valuable as we did.

Zhe Liang
Tongji University

Wanpracha Art Chaovalitwongse
University of Washington

Leyuan Shi
University of Wisconsin—Madison

MATLAB® is a registered trademark of The MathWorks, Inc. For product information, please contact:

The MathWorks, Inc.
3 Apple Hill Drive
Natick, MA 01760-2098 USA
Tel: 508-647-7000
Fax: 508-647-7001
E-mail: info@mathworks.com
Web: www.mathworks.com

Editors

Zhe Liang is a professor in the Department of Management Science in the School of Economics and Management, Tongji University, Shanghai, China. He received his BEng from the Department of Computer Engineering at the National University of Singapore in 2001, and his PhD from the Department of Industrial and Systems Engineering of Rutgers University, Newark, New Jersey, in 2011. His research interests are related to the design and implementation of exact and heuristic algorithms for large-scale combinatorial optimization problems in supply chain management, transportation, and telecommunication. He has more than 20 cited publications, including papers that appeared in journals such as *Transportation Science, INFORMS Journal on Computing, Transportation Research Part B,* and others. He has been funded by the National Science Foundation of China, Ministry of Education of China, and Ministry of Science and Technology of China.

Wanpracha Art Chaovalitwongse is a professor in the Departments of Industrial & Systems Engineering and Radiology (joint) at the University of Washington (UW), Seattle, Washington. He also serves as associate director of the Integrated Brain Imaging Center at UW Medical Center. Before moving to Seattle, he worked as Visiting Associate Professor in the Department of Operations Research & Financial Engineering at Princeton University, Princeton, New Jersey, in 2011. From 2005 to 2011, he was on the faculty in the Department of Industrial & Systems Engineering at Rutgers University, Newark, New Jersey. Before working in academia, he worked at the Corporate Strategic Research laboratory, ExxonMobil Research & Engineering, where he managed research in developing efficient mathematical models and novel statistical data analyses for upstream oil exploration and downstream business operations in multicontinent oil transportation. He received MS and PhD degrees in Industrial & Systems Engineering from the University of Florida, Gainesville, Florida, in 2000 and 2003. His research group conducts basic computational science, applied, and translational research at the interface of engineering, medicine, and other emerging disciplines with emphasis on computational neuroscience, computational biology, and logistics optimization. He has received more than $9M in research support from the National Science Foundation, the National Institutes of Health, the Institute of Museum and Library Services, Cisco—Academic Research & Technology Initiatives, and the New Jersey Schools Development Authority. He holds three patents of novel optimization techniques adopted in the development of seizure prediction system, which have been licensed to Optima Neuroscience, Inc. His academic honors include 2003 Excellence in Research from the University of Florida, 2006 NSF CAREER Award, 2007 Notable

Alumni of King Mongut's Institute of Technology at Ladkrabang, Bangkok, Thailand, 2004 and 2008 (2-time winner) of the William Pierskalla Best Paper Award by the Institute for Operations Research and the Management Sciences (INFORMS), 2009 Outstanding Service Award by the Association of Thai Professionals in America and Canada (ATPAC), 2010 Rutgers Presidential Fellowship for Teaching Excellence, 2014 Finalist of the UW College of Engineering Faculty Innovator Award, and several other best student paper awards with his PhD students. He has been invited to give lectures at top universities around the world, including MIT, Princeton University, University of Michigan, University of Warwick, University of London, Peking University, Chinese Academy of Sciences, Seoul National University, and KAIST. He currently serves as area editor, associate editor, and editorial board member of 10 leading international journals. He has edited 4 books and published more than 130 research articles including 70+ journal papers. In addition to research, he been actively involved with management and recently appointed as the president of the Association of Thai Professionals in America and Canada (ATPAC), which is a nonprofit organization in the United States with the main goal of promoting the advancement of scientific knowledge, technology, and education in Thailand. ATPAC works closely with Thailand's Ministry of Science and Technology and the Office of the Higher Education Commission.

Leyuan Shi is a professor at the Department of Industrial Engineering and Management at Peking University, Beijing, China. She received her PhD degree in applied mathematics from Harvard University, Cambridge, Massachusetts, in 1992, her MS degree in engineering from Harvard University in 1990, her MS degree in applied mathematics from Tsinghua University, Beijing, China, in 1985, and her BS degree in mathematics from Nanjing Normal University, Nanjing, China, in 1982. Her research is devoted to the theory and development of large-scale optimization algorithms, discrete event simulation methodology, and modeling and analysis of discrete dynamic systems, with applications to complex systems such as supply chain networks, manufacturing systems, communication networks, and financial engineering. She has published two books: *Nested Partitions Method, Theory and Applications* (2008) and *Modeling, Control and Optimization of Complex Systems* (2002), both published by Springer US. She is currently an editor of *IEEE Transactions on Automation Science and Engineering* and an associate editor of *Discrete Event Dynamic Systems*. Dr. Shi is a member of INFORMS and a fellow of IEEE.

Contributors

Paveena Chaovalitwongse
Department of Industrial
  Engineering
Chulalongkorn University
Bangkok, Thailand

Wanpracha Art Chaovalitwongse
Departments of Industrial and
  Systems Engineering and
  Radiology
University of Washington
Seattle, Washington

Weiwei Chen
Department of Supply Chain
  Management and Marketing
  Sciences
Rutgers Business School
Rutgers University
Newark and New Brunswick,
  New Jersey

Hui Dong
Amazon Corporate LLC
Seattle, Washington

Wenjuan Fan
School of Management
Hefei University of Technology
Hefei, China

and

Department of Computer Science
North Carolina State University
Raleigh, North Carolina

Kangbok Lee
Department of Business and
  Economics
School of Business and Information
  Systems
York College
The City University of New York
Jamaica, New York

Lei Lei
Department of Supply Chain
  Management
Rutgers Business School
Rutgers University
Newark and New Brunswick,
  New Jersey

Zhe Liang
School of Economics and
  Management
Tongji University
Shanghai, China

Rujing Liu
Praxair (China) Investment Co. Ltd.
Shanghai, China

Xinbao Liu
School of Management
Hefei University of Technology
and
Key Laboratory of Process
  Optimization and Intelligent
  Decision-Making of Ministry of
  Education
Hefei, China

Kwankeaw Meesuptaweekoon
Department of Industrial
    Engineering
Chulalongkorn University
Bangkok, Thailand

Athanasios Migdalas
Division of Industrial Logistics
Department of Industrial
    Engineering
Luleå University of Technology,
Luleå, Sweden

and

Division of Transportation,
    Construction Management, and
    Regional Planning
Department of Civil Engineering
Aristotle University of Thessaloniki
Thessaloniki, Greece

Tan Miller
Global Supply Chain Management
    Program
College of Business Administration
Rider University
Lawrenceville, New Jersey

John E. Mitchell
Department of Mathematical
    Sciences
Rensselaer Polytechnic Institute
Troy, New York

Sarah G. Nurre
Department of Industrial
    Engineering
University of Arkansas
Fayetteville, Arkansas

Rosa Oppenheim
Department of Supply Chain
    Management
Rutgers Business School
Rutgers University
Newark and New Brunswick,
    New Jersey

Panos M. Pardalos
Center for Applied Optimization
Department of Industrial and
    Systems Engineering
University of Florida
Gainesville, Florida

Jun Pei
School of Management
Hefei University of Technology
Hefei, China

and

Center for Applied Optimization
Department of Industrial and
    Systems Engineering
University of Florida
Gainesville, Florida

Lian Qi
Department of Supply Chain
    Management
Rutgers Business School
Rutgers University
Newark and New Brunswick,
    New Jersey

Ying Qin
School of Economics and
    Management
Tongji University
Shanghai, China

Pakpoom Rungchawalnon
Department of Industrial
  Engineering
Chulalongkorn University
Bangkok, Thailand

Thomas C. Sharkey
Department of Industrial and
  Systems Engineering
Rensselaer Polytechnic Institute
Troy, New York

Junmin Shi
School of Management
New Jersey Institute of Technology
Newark, New Jersey

Leiyuan Shi
Department of Industrial and
  Systems Engineering
University of Wisconsin–Madison
Madison, Wisconsin

Shengbin Wang
Department of Marketing,
  Transportation, and Supply
  Chain
School of Business and Economics
North Carolina A&T State
  University
Greensboro, North Carolina

Shaozhong Xi
Shanghai Railway Bureau
Shanghai, China

Yusen Xia
J. Mack Robinson College of
  Business
Georgia State University
Atlanta, Georgia

Jian Yang
Department of Management Science
  and Information Systems
Rutgers Business School
Rutgers University
Newark and New Brunswick,
  New Jersey

Section I

Supply Chain Strategy and Coordination

1

Supply Chain Frameworks: A Constant in the Midst of Change

Tan Miller

CONTENTS

ABSTRACT Supply chain management in the early 21st century continues to evolve, expand, and globalize at a dizzying pace. Whether it is adapting and integrating new decision support tools and new technologies related to business analytics, the Internet, equipment and machinery, or other infrastructure, new challenges confront supply chain managers every day. In this environment of constant, almost overwhelming change, it is critical not to lose sight of the fact that the underlying principles and foundation tools of supply chain management remain valuable, and even more important than ever. These principles, foundation tools, and methods provide an anchor, a guide map, a "constant" for supply chain managers to call on as they confront dramatic changes in their everyday work lives. In this chapter, we review and illustrate how to utilize effectively one key "foundational" supply chain management tool and methodology: multiple time horizon, hierarchical "frameworks" to organize and manage supply chain planning and operations activities.

Historical Footnote Multiple time period, hierarchical supply chain planning (HSCP) frameworks and methodologies have existed for decades. For perspective, we note that this hierarchical approach dates back to at least the 1960s, specifically to Robert Anthony's seminal work on planning and control systems (Anthony, 1965). Anthony classified all managerial decisions into three broad categories consisting of (1) strategic planning, (2) management control, and (3) operational control. A number of authors (e.g., Ackoff, 1970) eventually termed the second category as tactical planning and the third category as operational planning and scheduling. In its initial stages, the focus of these hierarchical planning systems or frameworks was integrated production planning and scheduling among the strategic, tactical, and operational levels. These systems were known as hierarchical production planning (HPP) systems (see, e.g., Hax and Meal, 1975). This early HPP work on coordinated production planning and scheduling activities across multiple time horizons set the stage for current HSCP systems.

KEY WORDS: *hierarchical production planning, logistics planning, supply chain frameworks, supply chain management.*

1.1 Introduction: What Is a Hierarchical Supply Chain Planning Framework?*

There are major organizational issues, systems and infrastructure considerations, methodology issues, and numerous other problem dimensions to evaluate in formulating a firm's logistics and supply chain network planning approach. From all perspectives, effective supply chain planning over multiple time horizons requires that a firm establish appropriate linkages across horizons and establish points of intersections between these horizons. To facilitate a planning system that possesses these appropriate linkages, a firm must have an overall framework that guides how different planning horizons and planning components fit together.

Figure 1.1 presents a general framework for hierarchical supply chain planning (HSCP) that defines three levels: strategic, tactical, and operations. As Figure 1.1 illustrates, strategic planning activities focus on a horizon of approximately 2 or more years into the future, whereas tactical and operational activities focus on plans and schedules for 12–24 months, and 1–18 months in advance, respectively. At the strategic level, a firm must address such key issues as overall corporate objectives, market share and profitability goals, business and product mix targets, and so on. Planning decisions on overall corporate objectives drive strategic supply chain decisions. For example, market share and business or product mix objectives will strongly influence manufacturing capacity strategies.

At the strategic manufacturing planning level, the firm must address such issues as planned production capacity levels for the next 3 years and beyond, the number of facilities it plans to operate and their locations,

---

* Much of the material presented in this introductory section first appeared in Miller (2009).

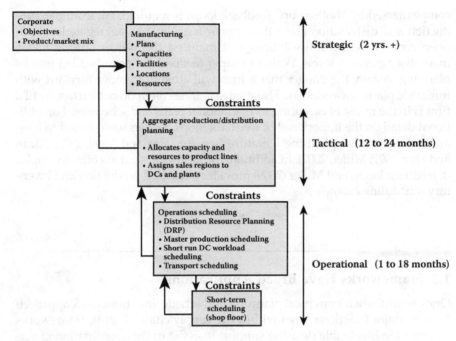

FIGURE 1.1
Hierarchical supply chain planning framework.

the resources the firm will assign to its manufacturing operations, and numerous other important long-term decisions. Decisions made at the strategic level place constraints on the tactical planning level. At the tactical level, typical planning activities include the allocation of capacity and resources to product lines for the next 12–18 months, aggregate planning of workforce levels, the development or fine-tuning of distribution plans, and numerous other activities. Within the constraints of the firm's manufacturing and distribution infrastructure (an infrastructure determined by previous strategic decisions), managers make tactical (e.g., annual) planning decisions designed to optimize the use of the existing infrastructure. Planning decisions carried out at the tactical level impose constraints on operational planning and scheduling decisions. At this level, activities such as distribution resource planning, rough cut capacity planning, master production scheduling, and shop floor control scheduling decisions occur.

Feedback Loops. The feedback loops from the operational level to the tactical level and from the tactical level to the strategic level represent one of the most important characteristics of the supply chain planning system illustrated in Figure 1.1. To ensure appropriate linkages and alignment between levels, a closed-loop system that employs a "top down" planning approach complemented by "bottom up" feedback loops is required. For example, production and distribution plans that appear feasible at an aggregate level can often contain hidden infeasibilities that manifest themselves only at lower, more disaggregated levels. Without proper feedback loops imbedded into its planning system, the danger that a firm will attempt to move forward with infeasible plans always exists. These infeasibilities often do not surface until a firm is in the midst of executing its operational plans and schedules. For additional detail on the importance of feedback loops, readers are referred to hierarchical production planning literature (see, e.g., Hax and Meal, 1975; Bitran and Hax, 1977; Miller, 2002). In addition, the following sections offer examples of feedback loops, and Miller (2009) provides a detailed production and inventory scheduling example.

1.2 Frameworks Have Broad Applications

One can utilize a hierarchical planning and scheduling framework approach for any major functional area within the supply chain. That is, frameworks can provide invaluable decision support to assist in the coordination of virtually all short-, medium-, and long-run planning and scheduling activities. As noted, the application of this methodological approach to supply chain management originated in the area of production planning and scheduling. However, transportation, inventory management, demand management, and warehouse operations represent just a few examples of major functional areas that can benefit from a hierarchical perspective (see, e.g., Miller, 2002). Further, frameworks can also be applied to more general decision support activities such as performance measurement. To illustrate the broad applicability of frameworks, we will now review the application of hierarchical frameworks to three diverse activities: warehouse operations, inventory management, and performance measurement.

1.3 Warehouse Operations

The warehouse planning process begins at the network-wide strategic planning level. At this level, a firm must determine how warehouse operations fit into its overall strategic plan, and in particular, what is the mission of each warehouse on its network, as well as on the overall network itself. Figure 1.2 provides a high-level overview of this hierarchical planning process that begins at the strategic level, while Figure 1.3 highlights selected decisions that take place at each level.

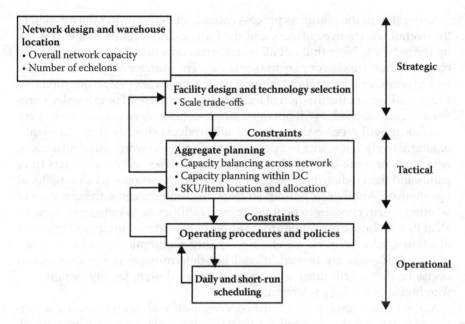

FIGURE 1.2
Hierarchical warehouse planning.

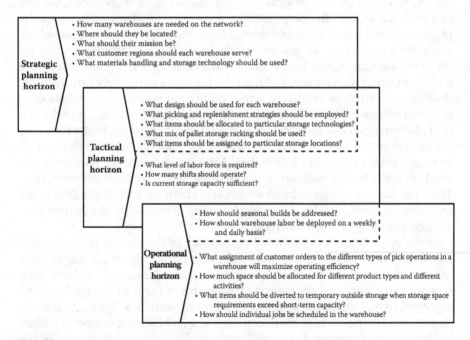

FIGURE 1.3
Framework of how warehousing decisions fit into a planning hierarchy.

A first step in the planning process consists of determining the mission of the overall warehouse network and the individual locations that will make up the network. Note that not all warehouses on a network will necessarily have the same mission or play the same role. The number of warehouse echelons to establish represents a common strategic network design question that heavily influences the mission of individual warehouses. For example, a firm must decide whether it will operate a single-echelon network in which every warehouse will receive shipments of all products directly from all plants, or alternatively if it wants to operate a multi-echelon warehouse network in which one or more first-echelon, central warehouses receive products from plants and then redistribute some or all products to second-echelon regional warehouses. Another important strategic decision concerns the question of whether a firm chooses to operate its own facilities or to outsource some or all of its warehouse operations to third-party providers. Finally, as Figure 1.2 illustrates, total network warehouse capacity requirements and economies of scale trade-offs are two additional key determinants of the interrelated decisions that a firm must address on network design, facility design, and warehouse technology selection.

At the tactical level, a firm must concern itself with such planning activities as balancing the demand for warehousing capacity across its network and planning the most efficient and effective utilization of its capacity at each individual distribution center (DC). Capacity planning at the individual DC level can involve determining the overall labor workforce level and mix required to meet the projected demands over the planning horizon, the proper mix and use of available storage locations (e.g., type of racking where adjustable), and so on. In general, tactical warehouse planning focuses on the determination of how best to employ the existing network infrastructure (i.e., the existing warehouses and material handling equipment). In addition, decisions to purchase relatively minor additional warehousing assets (e.g., incremental material handling equipment, racking, etc.) will occur in the tactical planning process. However, major infrastructure issues that a firm cannot resolve at the tactical planning level (e.g., inadequate network capacity to meet forecast long-term warehouse throughput or storage requirements) must typically be fed back up to the strategic planning level for resolution. Thus, the efficacy of hierarchical warehouse planning and scheduling relies on feedback loops, similar, for example, to the dependency of effective production planning on such mechanisms.

At the operational level, a broad assortment of warehouse planning and scheduling activities take place on a regular basis. Figure 1.3 illustrates a sample of key decisions that operational schedulers must address. The scheduling of labor and short-term assignments of items to storage locations represent two of the major operational planning activities. Typically, it is the nonroutine components of these activities (e.g., addressing temporary labor or storage requirements that significantly exceed capacity) that require the most critical attention. It is also typically the "exceptions" or "nonroutine" requirements of operational planning and scheduling that planners must report or "feed back" to the tactical planning level. For example, when warehouse planners consistently find themselves having to schedule "unplanned" outside storage because of insufficient facility storage capacity, they should send this information to the tactical level for resolution. Perhaps the overall warehouse network is out of balance and requires realignment because excess storage capacity exists at certain warehouses, while other warehouses face the opposite situation. Alternatively, perhaps this storage capacity issue at one warehouse is not an imbalance issue, but rather is occurring regularly across the network and requires a total network solution. This represents just one simple example of the types of feedback loops that must exist between the operational and tactical warehouse planning levels.

In summary, firms can improve their warehouse operations by explicitly maintaining a warehouse management framework that identifies

- What specific decisions are made at what level, by whom, and how often
- How these decisions impact long run to short run operations
- The planning and scheduling decisions support (dss) tools and systems that are utilized at each level of the planning hierarchy and how the inputs and outputs of these dss tools are linked and coordinated

1.4 A Framework of How Inventory Decisions Fit into a Planning Hierarchy*

Similar to production or warehouse management decisions, inventory decisions span all three levels of the planning hierarchy (i.e., the strategic, tactical, and operational levels). In addition, in a fashion similar to the production and warehousing functions, inventory decisions made at higher levels impose constraints on decisions that occur at lower levels, while at the same time the ramifications of constraints imposed by higher level decisions must flow back or upwards to higher planning levels from lower planning levels. Figure 1.4 offers an overview of where some key inventory decisions fit into a planning hierarchy. Although this set of decisions does not represent a comprehensive list, it serves to illustrate how key inventory decisions span multiple planning horizons. For illustrative purposes, we now consider several of these decisions in more detail.

At the strategic planning level, inventory decisions revolve around such questions as determining the optimal level of inventory investment that a

---

* Much of the material presented in this section originally appeared in Miller (2002).

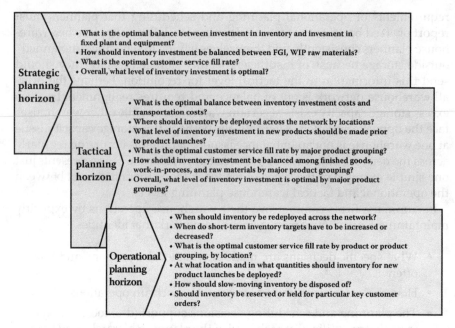

FIGURE 1.4
A hierarchical framework for inventory management.

firm should plan to maintain on an ongoing basis. This decision requires an evaluation of a number of intertwined trade-offs. For example, in many cases, manufacturing firms must consider what represents the best mix of investment in manufacturing capacity vis-à-vis total investment in inventory. As a general rule, for any given level of network-wide demand, as a firm increases its manufacturing capacity and flexibility, it decreases the level of inventory investment that it must make. This is particularly the case for make-to-stock firms. For example, if a firm has the capacity and flexibility to produce its entire product line during 1 week (i.e., cycle through its full set of products), it will require less finished goods inventory to fill customer orders than if the firm requires several weeks or months to cycle through its product line. In general, the more heavily a firm chooses to invest in fixed plant and equipment (i.e., production equipment), the greater its production capacity and flexibility. Thus, at the strategic level, a manufacturer must decide the best balance to maintain between fixed assets (plant and equipment) and inventory investment.

The inventory investment decision also requires that a firm consider the trade-offs between relative levels of investment in finished goods inventory versus raw materials and work-in-process components. In addition, at a very high level, a firm must develop a strategy on its planned customer service level fill rate. For example, a make-to-stock manufacturer cannot simply plan to have a 100% customer service line fill rate. This is not realistic. Thus, a manufacturer must put a service level strategy in place considering the trade-offs between some "acceptable" level of lost or unfulfilled demands versus the inventory investment associated with alternative service levels. Finally, such issues as the appropriate trade-offs between transportation costs and inventory investment costs also require consideration in the inventory management process. Often manufacturers must choose between using faster, more expensive transport services (e.g., air) between a "supplying" origin and a "receiving" destination and slower, less expensive transportation alternatives. A key question in such decisions is whether the savings in inventory investment requirements (i.e., in-transit and safety stock inventory at the destination) facilitated by a faster replenishment transit time outweigh the increased costs of the expedited transport service. Questions such as this often represent tactical rather than strategic issues; however, they play a prominent role in a firm's overall inventory investment approach.

At the tactical planning level, many of the same inventory issues found at the strategic level resurface, but at a greater level of granularity. For example, typically a firm's annual planning process includes an evaluation of the level of the total inventory investment that it will make during the next year. At the minimum, however, this process will usually include an analysis at the major product grouping level (e.g., brand or product family). In contrast, the strategic inventory planning process will not address this level of detail. Many of the same trade-offs considered at the strategic level again reappear in more detail (e.g., evaluating optimal service fill rates and the approximate mix of finished, in-process, and raw materials inventory at the product grouping level). Further, as previously noted, decisions implemented at the strategic level will impact options at the tactical level. For example, the manufacturing capacity and flexibility built into the firm's current infrastructure will heavily influence its annual inventory investment plan.

At the tactical level other more detailed decisions such as the deployment of inventory by location and by product grouping will also occur for the first time. An example of the type of feedback that can develop at this level would be if a firm's planners, when reviewing inventory requirements by major product grouping, determine that to meet planned service fill rates, they require a total inventory level exceeding the planned overall investment target. In addition, at this level, specific policies must be developed covering such potential issues as whether certain customers will receive priority in the event of temporary inventory shortages during the planning horizon. The development of guidelines for when inventory should be redeployed between locations on a firm's network because of shortages and/or imbalances represents another example of the types of policies developed at the tactical planning level.

At the operational planning and execution level, again many previous planning decisions made at higher levels are revisited at a more detailed product line level and in more detailed time increments. Here a firm's planners must ensure that all SKUs (i.e., stock keeping units, or items at unique locations) have inventory targets designed to deliver specific customer service fill rates. In addition, decisions regarding inventory targets by season, by month (perhaps even by week) must be made, and made within the guidelines or constraints established at higher planning levels. Inventory policies represent a vital area in which the existence of good feedback loops from the operational level to the tactical level can play an important role. If, for example, the guidelines regarding inventory redeployment between network locations in cases of inventory imbalances are not working effectively, this information must be communicated back to the tactical planning level so that a more effective approach can be developed and implemented.

In summary, the inventory management function shares the same critical need for coordination and feedback between hierarchical planning levels as do other functions such as warehouse operations and manufacturing. A firm that makes the effort to formally design and update on a regular basis an inventory management framework positions itself to plan its inventory investment and deployment efficiently and effectively.

1.5 A Hierarchical Supply Chain Performance Measurement System

This section presents a three-level hierarchical performance measurement framework that can link a wide array of functional areas and performance measures. The three levels of the hierarchy consist of the strategic, tactical, and operational levels, as found in traditional supply chain frameworks. However, the hierarchical levels have a different connotation in this performance measurement system (PMS). As discussed, in traditional frameworks, the dimension of "time" or planning horizon differentiates the hierarchy levels. In our PMS framework, it is the "scale" of an operation or activity that a particular performance measure monitors that determines its place in the hierarchy. Also within the hierarchy of the firm, the level of a functional unit that a measure monitors determines where it (i.e., the measure) falls in the performance measurement hierarchy.

Within each of the three levels of the measurement system, we further differentiate performance measures into two categories: (1) external measures and (2) internal measures. External measures focus on the effectiveness of an activity or function while internal measures evaluate the efficiency or productivity of an activity or function. In particular, external measures evaluate the effectiveness of flows and links across a supply chain, while internal measures evaluate the cost or efficiency of a function or organization in producing its outputs and services. Two typical external performance measures are order and line item fill rates on customer orders. When a customer (e.g., a mass merchandiser) places an order to a supplier (e.g., a manufacturer), the order and line item fill calculations measure whether the supplier delivers the total order and the individual line items on time and complete as ordered. These measures do not evaluate the supplier's cost in delivering the order. Thus, if a supplier had to deliver an order by expensive air freight rather than normal surface transportation because of inventory shortages or other problems, the order and line fill calculations would not capture this. If the shipment arrived on time and complete by air transport, the order fill and line fill measures would identify this particular shipment as a successful fulfillment of a customer order (i.e., the manufacturer was "effective" in delivering the order on time and in full to the customer). In contrast, internal performance measures focusing on efficiency would evaluate this order fill example as an "inefficient" effort. "Distribution cost per case" and "freight cost per lb." represent two common internal measures, and this air freight delivery would increase the firm's average distribution cost per case and freight cost per pound.

Figure 1.5 depicts an integrated hierarchical supply chain performance measurement framework. Note that the framework in Figure 1.5 spans an entire supply chain from the echelon of the raw material and component suppliers, to the echelons of the plants and distribution centers of a manufacturing firm, to the echelon(s) of the locations of the customers of a manufacturer, and finally to the retail consumer. One can adapt the performance

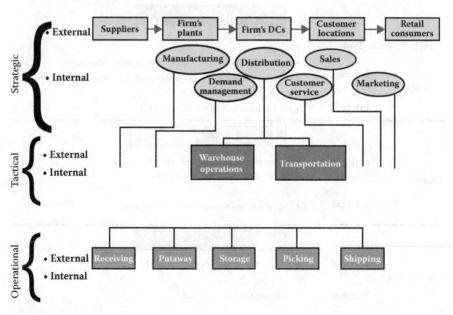

FIGURE 1.5
A hierarchical supply chain performance measurement framework.

measurement framework in Figure 1.5 to as many or as few echelons as may exist on a supply chain. For illustrative purposes, we now consider this framework from the more limited viewpoint of a single firm. To measure and monitor both the effectiveness and the efficiency of each of its major functions, a firm will want to have a few key high level measures, both internal and external, for each function. To describe this framework, we briefly focus on the distribution organization in particular.

As noted, at the strategic level, a few key indicators will measure the overall performance of the distribution organization. At the tactical level, the performance of each major subfunction of a functional area is measured. In this example, warehouse operations and transportation represent the two primary components of distribution. Therefore, we will need a few internal and external performance measures for each of these subfunctions. At the operational level of this framework, the performance of each key subfunction (of each function of the tactical level) is monitored. In this example, Figure 1.5 illustrates that the warehouse operations component has five major subfunctions: receiving, putaway, storage, picking, and shipping. Thus, each of these five areas would require both external and internal performance measures. Similarly, the major subfunctions of transportation would each require their own measures.

Figure 1.6 provides additional insight on how this hierarchical performance measurement framework works. This figure displays sample external

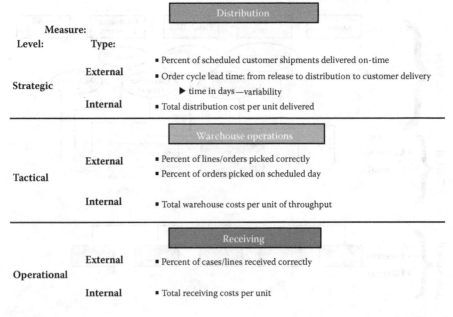

FIGURE 1.6
Illustrative hierarchical performance measures for the distribution function.

and internal performance measures for a distribution organization at each level of the hierarchy. The measures of (1) percent of scheduled customer shipments delivered on time and (2) the average and variance (or standard deviation) of order cycle lead time represent strategic external measures. These indicators are external because they measure outputs and/or services that flow across the supply chain. In this case, they provide a measure of how effectively the distribution function serves its customers at the next echelon of the supply chain. These measures are classified as strategic in this hierarchy because they evaluate one of the key missions of a distribution function: to distribute product and/or services to its customers in a timely manner. The other strategic indicator shown in Figure 1.6 is the total distribution cost per unit delivered (e.g., per case delivered). We classify this measure as internal because it has an "inward" focus; namely, it evaluates how efficiently the distribution organization performs its function. This internal measure falls into the strategic level because it evaluates a major function (distribution) at a summary level (i.e., the distribution cost per case delivered provides a high level view of overall distribution cost and effectiveness).

The performance measure examples shown in Figure 1.6 for warehouse operations (at the tactical level) and warehouse receiving operations (at the operational level) follow a similar theme. The percent of lines or orders picked correctly and the percent of orders picked on the scheduled day represent external measures because they evaluate the impact of warehouse operations across the supply chain. Specifically, when a warehouse picks an order correctly, it contributes to the ultimate successful delivery of products to a customer that has placed an order. In contrast, an incorrect pick will result in an unsuccessful delivery to a customer. Similarly, when a warehouse picks an order on time (i.e., on the scheduled day), this contributes toward a successful on-time delivery of products to a customer. The third tactical measure shown in Figure 1.6, total warehouse costs per unit of throughput, represents an internal measure, as it offers a summary view of the internal cost (and efficiency) of the warehouse operation.

At the operational level, the warehouse receiving function uses the percent of cases (or lines) received correctly (i.e., accurately) as an external performance measure. We categorize this measure as external because the accuracy with which this function receives inbound shipments will impact the next stage of the supply chain. For example, suppose that the receiving area miscodes an inbound receipt as product A, when in fact it received product B on a delivery. If this error remains undetected, the shipment will then be put into inventory classified as product A and at some future point, could be picked and delivered to a customer who ordered product A. In contrast, total receiving cost per unit has an internal orientation and will be of most immediate concern to receiving and warehouse personnel.

Benefits of a Hierarchical Performance Measurement Framework. Employing a hierarchical performance measurement framework offers a number of important benefits. First of all, this approach provides a unified framework for aggregating performance measures across a company. By implementing a hierarchical measurement approach, a firm positions itself to organize its key performance measurements into a structure that leads to a relatively few, high-level, strategic measures that monitor overall firm performance. At the same time, this structure facilitates having many other performance measures that monitor smaller components of a firm's operation, yet that align with overall firm objectives. In particular, a hierarchical measurement system allows both large and small functional areas within a firm to develop and maintain their own measures and to contribute to and be part of an overall measurement system.

This hierarchical approach thus helps to keep measures both "simple" and "meaningful" because each function and subfunction at each level can focus on a few key performance measures.

Finally, a hierarchical performance measurement system can also contribute toward aligning the collective activities of a firm to meet a desired mission and set of objectives. For example, if a firm has a comprehensive measurement system in place that covers most or all of its major functional areas and activities, managers can view the system in its entirety to identify any potential misaligned activities or objectives.

In closing this section, we emphasize again that the PMS framework differs markedly from the other frameworks presented in that "time or planning horizon" does not distinguish the levels of the hierarchy. As described, "scale or magnitude" differentiates one level of the PMS hierarchy from another, and this illustrates the flexibility of the framework paradigm. In short, frameworks can be applied to a highly diverse set of supply chain functions and activities.

1.6 Summary and Conclusions

We prefaced this chapter by acknowledging that supply chain professionals, as well as their colleagues in all other parts of the business world, must manage today in an unprecedented era of technological advancement and rapid globalization. Competitive pressures continue to escalate for firms as customers' physical and informational access to competitive suppliers around the world increase through developments such as enhanced transportation and distribution infrastructure, the Internet, and business analytics. In this chapter, we have made the point that in the midst of this dynamically changing world, supply chain professionals must more than ever utilize and call on well tested operating principles and methods.

HSCP frameworks represent one of these fundamental methods that practitioners should employ to help manage their operations. We explored three different frameworks for warehouse operations, inventory management, and performance metrics respectively in this chapter. Although each framework was different, they were also similar in that each provided a planning platform to coordinate the activities of a major function from the long-run strategic horizon to the short-run operating horizon. Thus, managers can use these hierarchical planning and scheduling systems to help organize and align the activities of major functional areas, and an entire supply chain over a multilevel operating horizon.

Finally, we note that although this chapter served to introduce the concept of HSCP frameworks, we did not review in detail some of the key "mechanics" and decision support tools of these systems, nor did we describe actual real-world implementations. Readers interested in exploring HSCPs further are referred to (Liberatore and Miller, 1998; Miller, 2002; Miller and Liberatore, 2011) for details on implementations and related strategies, as well as examples of additional frameworks.

References

Ackoff, R. *A Concept of Corporate Planning*. New York: Wiley Interscience, 1970.

Anthony, R. *Planning and Control Systems: A Framework for Analysis*. Boston: Graduate School of Business Administration, Harvard University, 1965.

Bitran, G. and A. Hax. On the design of hierarchical production planning systems. *Decision Sciences*, 8: 28–55, 1977.

Hax, A. and H. Meal. Hierarchical integration of production planning and scheduling. In M. A. Geisler (ed.), *TIMS Studies in Management Science, Vol. 1: Logistics*. Amsterdam: Elsevier, 53–69, 1975.

Liberatore, M. and T. Miller. A framework for integrating activity based costing and the balanced scorecard into the logistics strategy development and monitoring process. *The Journal of Business Logistics*, 19(2): 131–154, 1998.

Miller, T. *Hierarchical Operations and Supply Chain Management*, 2nd ed. New York: Springer Science+Business Media, 2002.

Miller, T. Notes on using optimization and DSS techniques to support supply chain and logistics operations. In W. Chaovalitwongse, K. Fueman, and P. Pardalos, *Optimization and Logistics Challenges in the Enterprise*. New York: Springer Science+Business Media, 2009.

Miller, T. and M. Liberatore. A practical framework for strategic planning. *Supply Chain Management Review*, March/April: 38–43, 2011.

2

Future Research on Multiobjective Coordinated Scheduling Problems for Discrete Manufacturing Enterprises in Supply Chain Environments

Jun Pei, Xinbao Liu, Wenjuan Fan,
Athanasios Migdalas, and Panos M. Pardalos

CONTENTS

ABSTRACT   New challenges are arising in supply chain management with growing competition in global markets. To address these challenges, supply chain partners must cooperate with each other much more closely to decrease costs and increase their competiveness. In particular, cooperation between partners in discrete manufacturing enterprises is of special importance. This chapter focuses on the coordinated scheduling problems of discrete manufacturing enterprises in a supply chain environment. It provides a comprehensive review from various perspectives, identifies interesting problems from the existing literature, and proposes prospective research directions. The research methodology mainly applied for solving these problems is briefly discussed.

KEY WORDS: *coordinated scheduling, supply chain, discrete manufacturing.*

2.1 Introduction

In the era of E-commerce, discrete manufacturing enterprises (DMEs) are characterized by multitype, small-batch, and flexible production patterns because of personalized and timing requirements of customers. DMEs are vital for the economy and the competition among them is increasingly becoming a competition in the supply chain. However, the matching, coordination, and optimization between the production of most DMEs and the supplying and distribution of up- and downstream partner enterprises are not factored into the production and scheduling plans of the DMEs, resulting in an increase in the total costs of the supply chain and a weakening of their competitiveness. In the existing literature, the research work on scheduling problems in a supply chain environment is limited to the production plans, the established models are generally too simple, the objectives and constraint conditions are too idealized, and the production features of DMEs are not taken into account, which is far from the real-life situations (Frederix, 2001). Thus, the research on the supply chain problems for DMEs is of significant theoretical and practical importance.

Characteristic features of the production systems of DMEs are complicated products, multitype material, uncertainty, and focus on software. Indeed, (1) the products of DMEs tend to be complicated because they may be composed of many components, and they usually have relatively fixed matching relationships between the products' structures and the components; (2) DMEs' products are diverse, which requires multiple types of production

materials; (3) in DMEs, connection and management among different departments or different production processes frequently bring high uncertainty
into production; (4) the productivity of DMEs does not depend on the hard-
ware but on the software such as management, unlike in continuous manufacturing enterprises.

The production scheduling patterns in a supply chain environment have
some new features that are different from the traditional ones (Hall and
Potts, 2003).

1. There is a greater optimization space for coordinated scheduling patterns in a supply chain environment, with more numerous,
   broader, and more complicated involved departments. The scheduling decisions in a supply chain are made in the workshop and
   the functional departments in the form of a multitype and small-
   batch approach not only in the complicated production processes
   for each job-machine assignment and the job processing orders on
   each machine, but also for the integration of each node in the supply
   chain network, to optimize globally the production activities of each
   functional department, such as transportation, storage, and so forth,
   so as to enhance the competitiveness of the entire supply chain.

2. The multiobjective feature of production scheduling in a supply
   chain environment differs from traditional multiobjective production scheduling problems because it focuses on the "benefit balance"
   among the involved nodes in the supply chain network. The corresponding problems are not limited to the traditional production area
   but also bear a close relationship with materials, finished products,
   storage, transportation, and other areas. In view of this, the main
   purpose of coordinated scheduling in a supply chain environment
   is to realize global optimization of the supply chain so as to enhance
   the competitiveness of the entire supply chain and not restrict the
   goal to the optimization of some objective for a single production
   node or functional department, such as maximizing machine utilization and in-time delivery ratio, minimizing the completion time,
   and so forth. Moreover, its purpose is not the optimization of the
   aggregate of the objectives of the nodes, but of the "benefit balance"
   across multiple nodes in the supply chain network. This pattern,
   besides the traditional optimization objectives in production systems, also involves the transportation and storage problems among
   multiple nodes, such as how to maximize the on-time arrival rate,
   minimize production storage and transportation costs, or maximize
   the output value of the entire supply chain.

Traditional production scheduling problems address more academic
issues such as mathematical modeling theory and methods, complexity

theory, control theory, combinatorial optimization theory and methods, artificial intelligence, computer technology, and so on, whereas the coordinated scheduling problems in a supply chain environment address more managerial issues, because besides scientific methodology, effective communication and coordination among all nodes are required. That is, the collaborative management mechanism and supply chain operation cost is vital for supply chain production scheduling and constitutes the basis for production scheduling theory and method implementation.

Future research should aim at the new features of a DME production scheduling pattern in a supply chain environment, at studies of the existing supply chain production scheduling collaboration management mechanism, multiobjective modeling and solution methods for typical problems, and at the development of evaluation methods. It should include research on (1) the new features of scheduling problems in a supply chain environment; (2) the collaboration mechanism of enterprise scheduling in a supply chain environment; (3) the DME production scheduling models in a supply chain environment; (4) the solution methods and simulation experiments for DME production scheduling models in a supply chain environment; and (5) case studies.

2.2 Literature Review

Next we review the multiobjective production scheduling problems in a supply chain environment from the aspects of coordination mechanism of the production system, production plans, production (machine) scheduling, multiobjective production problem-solving techniques, and multiobjective evolutionary algorithm evaluation methods.

2.2.1 The Research Status of Coordination Mechanisms for Production Systems in a Supply Chain Environment

The coordination mechanisms for production systems generally include the production scheduling management mechanisms in a supply chain environment, the information communication mechanism of production scheduling in a supply chain environment, and the benefit distribution mechanism among the supply and demand nodes in the up-/downstream supply chain.

The production scheduling management mechanisms in a supply chain environment include the joint replenishment management, joint inventory management, joint production, production and transportation collaborative optimization, and production outsourcing mechanisms, among others. The joint replenishment management mechanism has many submechanisms such as rapid-reaction, accurate-response, backup agreement, and quality–flexibility mechanisms. The rapid-reaction mechanism was first proposed by Iyer and Bergen (1997). Sahin et al. (2008) used the method of frozen orders to determine fixed deliveries in a certain period of time, with the remaining orders decided when the next period approached, so as to realize the effect of rapid reaction. In the accurate-response mechanism, the manufacturers in a supply chain designate a two-stage materials production for the material producers to realize the flexibility of production supply, considering the random requirements of the market (Fisher and Raman, 1996). The joint inventory management mechanism in a supply chain environment requires that each entity should consider not only its own production and inventory situations, but also the other partners' when adopting the inventory management policy, with the result that the entire supply chain can realize inventory optimization. The joint production management mechanism in a supply chain environment includes joint production of parallel enterprises and up-/downstream enterprises. Frederix (2001) proposed a method of expanding enterprise plan for DMEs, in which he considered that the core enterprise was in charge of the coordination and scheduling of multiple partners' production in the supply chain, so as to realize efficiency optimization of the whole supply chain's resource. Alvarez (2007) also studied the problem of expanding the enterprises' plan, in which each producer was considered as an independent agent, the coordination and control task of which was not taken by the core enterprise only. Kanyalkar and Adil (2005) considered the problem of production coordination management in a supply chain that has multiple production equipment locations, with alternative production capability supplying to multiple points of sale and suppliers that are dynamically determined. Sawik (2009) developed a coordination mechanism and two mixed integer programming methods with respect to the problem of multiple suppliers and producers in a supply chain to realize the synchronization of materials supplying and minimize the total inventory and other costs of the entire supply chain. Lee and Chen (2001) studied the overall coordination scheduling problem of production and transportation considering the transportation capability and time, based on the research of machine scheduling, and also discussed the complexities of a variety of scheduling problems related to this problem. Chang and Lee (2004) further studied such problems taking into account the differences of storage space of each product. Li et al. (2005) considered the transportation coordination problem when customers are distributed in different locations. The production outsourcing mechanism in a supply chain environment is subject to the coordination mechanism applied in the situation when the production capacity or production types of core enterprise are insufficient and the production is outsourced to other enterprises. The information communication and sharing mechanism in a supply chain environment is the basis of a unified, coordinated, and efficient operation of the multiple nodes and the entire supply chain. Jayaraman and Pirkul (2001) pointed out that information on the production process is the biggest concern. The benefit

24 *Supply Chain Management and Logistics*

distribution mechanism among supplying and demanding nodes of the up-/ downstream supply chain is the focal issue in the cooperation and dissention of supply chain enterprises. Leng and Parlar (2009) studied a two-node supply chain with producer and distributor, using Nash equilibrium theory to construct a linear benefit distribution mechanism.

Given the research status on the production scheduling collaborative mechanism in a supply chain environment, studies on the supply chain production coordination mechanism have a direct relationship to laws, credit, and even management systems applicable to the enterprises.

2.2.2 The Research Status of Joint Production, Production Inventory, and Production Transportation Plans in a Supply Chain Environment

Since the concept of a supply chain was first proposed, the production plan in a supply chain environment has been a hot area in academic research. Taking the joint production plan, for example, Kanyalkar and Adil (2005) studied the collaborative production plan problem in a multifactory, multipoint of sale environment and proposed a linear programming model to realize unified plan coordination of production time and production capacity. Alvarez (2007) discussed the coordination plan and scheduling problem of expanding enterprises and the development level of this research area. In research work on production inventory plans, Jayaraman and Pirkul (2001) studied the strategy level decision-making problems of multiproduction, multisupplier, multifactory, and multicustomer enterprises. The authors developed a solution method based on heuristics for the coordination plan. In addition, Xie et al. (2006) considered the inventory factor in the coordination problem in a supply chain environment. In research work on production transportation plans, Liang (2007) studied the comprehensive planning problem of production and transportation in a supply chain environment and constructed an interactive fuzzy multiobjective linear programming model. Aliev et al. (2007) proposed a fuzzy genetic algorithm to integrate production and transportation in production planning of a supply chain. There are also related works in Demirli and Yimer (2006), Rizk et al. (2006), Chern and Hsieh (2007), and Selim et al. (2008). Torabi and Hassini (2008) conducted research on the coordination plan in a supply chain in the form of purchasing–producing–distributing.

There is extensive and deep related work on the production planning research in a supply chain environment. Research on constraints and objectives involves the production cost, production capacity, setting time, setting cost, cleaning time, transportation time, transportation vehicle space, transportation cost, transportation capacity, inventory level, inventory cost, warehouse cost, no-delivery cost, inventory service level, replenishment cost, replenishment time, order flexibility, delivery time, dynamics, and randomness. The modeling methodologies used by the researchers include the mathematical programming models such as the linear/nonlinear, two/ multistage, random/nonrandom, single-objective/multiobjective, fuzzy/ nonfuzzy, and mixed integer/integer programming models, as well as intelligent algorithms such as genetic algorithm, tabu search algorithm, ant colony optimization algorithm, and many heuristics algorithms. In view of this, research on production planning in a supply chain environment is very comprehensive, in contrast to research on production scheduling in a supply chain environment, which is still limited. Mula et al. (2010) conducted a literature review on production and transportation planning problems in a supply chain environment and found that more than 90% of all the 127 investigated related work concerned the problem of planning level, not scheduling level.

2.2.3 The Research Status of Production Scheduling in a Supply Chain Environment

In 1996, Rowe et al. (1996) first proposed the concept of logistics scheduling, which introduced queuing theory into the supply chain management area. Afterward, Hall and Potts (2003) published the first paper on supply chain scheduling. In 2004, Hall gave a speech on supply chain scheduling and introduced the further research results on this topic (Agnetis et al., 2006; Dawande et al., 2006). Since then, research in this area has been wide ranging. The supply chain scheduling problem is a collaborative optimization problem with the optimization objective of minimizing the sum of production and transportation costs, integrating production scheduling, and splitting shipment. Zegordi et al. (2010) studied a two-stage scheduling problem in the supply chain, in which production scheduling was divided into production and transportation stages. The first stage involved multiple suppliers with different production speeds, and the second stage had multiple vehicles with different speeds and transportation capacities. The authors constructed the mathematical model for the problem as a mixed integer programming problem and used a new genetic algorithm to solve it. Mazdeh et al. (2011) applied the branch and bound algorithm in solving the collaborative optimization problem of single-machine production scheduling and batches delivery, and the optimization objective was to minimize the total cost of production scheduling and batch delivery. In addition to the aforementioned research work on the supply chain scheduling problem involving the coordination scheduling of production and transportation, some authors studied the production collaborative scheduling problem with multiple factories or multiple stages. For example, Agnetis et al. (2006) investigated the multifactory collaborative scheduling problem. Their article was not about the scheduling of parallel enterprises but about the up- and downstream enterprises, that is, multistage or multilevel supply chain scheduling. The authors considered the benefit of both suppliers and manufacturers through an intermediate storage buffer to realize optimization of production scheduling for both parties. Other issues of concern in production scheduling in a supply chain environment are the collaborative optimization problems of production scheduling and inventory and the collaborative optimization problem of production scheduling, inventory, and transportation. Selvarajah and Steiner (2006) took the transportation and inventory costs into account when studying the production scheduling problem in a supply chain. Kaminsky and Kaya (2008) introduced the problem of inventory location selection when studying the job scheduling problem in a supply chain environment. We also investigated coordinated scheduling and transportation problems in an aluminum supply chain and developed an effective intelligent algorithm and heuristic algorithm, respectively (Pei et al., 2014a,b).

In conclusion, research on coordination production in a supply chain is mainly on the strategy or planning level but limited on the scheduling level, and all of them do not consider the DME's supply chain scheduling problem with complex production conditions such as multitype, small-batch patterns and flexibility. There is still a lack of research on the supply chain scheduling problem from the perspective of multiple objectives, especially from the perspective of decision makers.

2.2.4 The Research Status on the Solving Techniques for Multiobjective Production Scheduling

The methods for solving multiobjective scheduling problems are generally classified into three groups: converting the multiobjective scheduling problem into a single-objective one by weighting the objectives and searching for the non-Pareto the Pareto solutions.

In research work on the first group of methods, the determined weights strategy is mainly used. Xia and Wu (2005) studied a triple-objective flexible job-shop scheduling problem (FJSP) in which the three optimization objectives were converted into a single one using the weighting method, and then a mixed algorithm of a tabu searching algorithm and particle swarm searching algorithm was used to solve the problem. Huang and Yang (2009) describe other studies using similar weighting strategies. In multiobjective problems, the dimensions of different objectives are usually different; thus some methods unified the dimensions of objectives before weighting them. Cardoen et al. (2009) studied the scheduling problem of doctors and patients in a hospital, where six objectives were taken into account, and the multiobjective scheduling problem was converted into a single-objective one by using the weight dimension conversion formula to each objective, and then the authors established the mixed integer linear programming model, based on which a deterministic algorithm and a heuristic algorithm were proposed to solve it. The second type of method is to find non-Pareto solutions. Low et al. (2006) studied a triple-objective mixed flowshop problem, the objective of which was to minimize average processing time, minimize average job tardiness, and minimize machine idle time. The authors proposed a mixed algorithm of simulated annealing algorithm and tabu search algorithm. Zhang et al. (2009) also studied a triple-objective mixed flowshop problem, and the difference was that the authors considered the problem from two parts, the first part of which used a particle swarm algorithm to determine the job assignment on each machine, and the second part was done by a simulated annealing algorithm to determine the processing order of the jobs on each machine. Related papers include Yagmahan and Yenisey (2008). The third type of method is to find Pareto solutions, which are mainly intelligent algorithms or mixed intelligent algorithms, and most of them consist of genetic algorithms. Some articles focused on how to find the most fitting Pareto solutions after obtaining the Pareto solution set. Pasupathy et al. (2006) studied the multiobjective permutation flowshop problem, the objective functions of which were to minimize the manufacturing cycle and minimize total processing time. The authors proposed a multiobjective genetic algorithm based on Pareto sorting, the so-called Pareto genetic algorithm, and also proposed the concept of nondominated sorting and crowding distance. Besides the genetic algorithm, there are some other algorithms also applied to such problems, for example, immune algorithm, particle swarm algorithm, and simulated annealing algorithm. Tavakkoli-Moghaddam et al. (2007) studied a multiobjective nonwaiting time flowshop problem, of which the objective functions included minimizing average completion time and minimizing weighted average tardiness. Behnamian et al. (2009) used a special method that combined the random weighting strategy and the finding Pareto solutions strategy to solve a multiobjective mixed flowshop problem in a real production scenario.

The research work described in the preceding text discussed the modeling and solving problems of a multiobjective scheduling problem in a single production unit. However, in a real production process, production, inventory, and transportation are tightly connected. Integrating these three key problems; establishing models and proposing solution methods; and in the end forming effective production scheduling plans, inventory strategy, and transportation management mechanisms hold great application prospects.

2.2.5 The Research Status of Multiobjective Evolutionary Algorithms Performance Evaluation

There have been some research results on the multiobjective evolutionary algorithm performance evaluation (Chinchuluun et al. 2008). Knowles (2002) conducted a comprehensive and systematic analysis on multiobjective evolutionary algorithms, which became the basis of future research. Deb et al. (2002) proposed two evolution indicators related to distance, which provided a new idea for evaluation indicator design of multiobjective evolutionary algorithms. However, these two indicators must be operated on the basis of a known Pareto-optimal solution set, but in real applications the Pareto-optimal solution set is often unknown. In view of this, it has very limited value for real-world applications.

The existing evaluation indicators of multiobjective evolutionary algorithms are insufficiently applicable to real problems. Moreover, the algorithm performance cannot be evaluated effectively unless the multiple indicators can interwork well together. However, there is still no recognized and effective solution for how to combine and apply multiple indicators.

2.2.6 Existing Problems

Here we propose some existing problems as follows:

1. The production system coordination mechanism in a supply chain environment is an important procedure for ensuring that production scheduling theories and methods in a supply chain are effectively implemented. Problems in this area include

   - How to coordinate the activities among and within enterprises, or among the functional departments such as factories
   - How to realize the global optimization of production scheduling in a supply chain
   - How to coordinate and manage the processes of supplying, production, sale, inventory, and so forth, so as to reduce the total costs and increase operation efficiency
   - How to coordinate the sharing and distributing problems of the benefit, risk, and costs among all functional departments in a supply chain
   - How to construct the information sharing platform to realize the global optimization of production scheduling in a supply chain

   Such problems are little discussed in the existing research materials, and the depth, range, and real applicability of the current research status cannot satisfy the practical requirements, and thus they need to be studied further.

2. There are some research results on joint production planning and control and on production inventory planning and production transportation planning. However, the research work hardly targeted to production scheduling, which is on the operational level in a supply chain. Thus, ways to conduct in-depth studies on the operational level (i.e., production scheduling level) are a significant direction for future research.

3. Research on production scheduling in a supply chain environment is still in the early stage, that is, the studied problems are oversimplified, and most of them contain only a few constraints, being far removed

from the practical production situation in industry. Thus, how to extend research on the specific problems to bring them much closer to the real situation is another future research topic of great significance.

4. Research on the production scheduling problems in a supply chain has focused mostly on model building, with little breakthrough on the solution methods. Most researchers used existing or simple algorithms, such as exact algorithms, heuristic algorithms, and so forth, to solve small-size problems. However, small-batch and multitype features are two of the most important characteristics of DMEs, and thus the problems are usually of a large size. Effective algorithms that can solve large-size problems in a reasonable time should be developed so as to apply production scheduling theory and methods in a practical DME production environment. Thus, further in-depth research on the solution techniques for large-size production scheduling problems should be conducted.

5. The multiobjective algorithm theory and methods based on the concept of Pareto optimality has become a hot topic in multiobjective algorithms in recent years. There are some results that have been applied well in many areas, and also some progress in traditional production scheduling problems. The production scheduling problems in a supply chain environment are mostly multiobjective problems, but there is still limited research on the algorithms for finding a Pareto-optimal solution set. Problems include

   • How to apply the multiobjective algorithms based on the concept of Pareto optimality in the multiobjective production scheduling problems in a supply chain

   • How to develop effective multiobjective evolutionary algorithms in view of such problems

   • How to combine existing knowledge, known scheduling rules, heuristics algorithms, and multiobjective evolutionary algorithms

   All of these problems need to be studied further.

6. There are a few distinct advantages in the real applicability of the evaluation indicators for multiobjective evolutionary algorithms based on the concept of Pareto optimality, but they also lack a unified evaluation framework. Moreover, there are no unified measurements and testing data for the production scheduling problems in a supply chain. These disadvantages decrease the accountability and fairness of the evaluation of the algorithm. Thus, how to design the measurement index with more practical operability, how to build a unified algorithm evaluation framework, and how to construct or generate testing data closer to the real situation are problems that urgently need to be studied.

2.3 Future Research Directions

2.3.1 Research on the New Features of Production Scheduling Problems in a Supply Chain Environment

There exist significant differences between the production scheduling problems in a supply chain environment and traditional production scheduling problems in a single workshop or single factory. Given the status of production scheduling management in up-/downstream DMEs, we need to analyze the new features and complexities of the DMEs' production scheduling problems in a supply chain environment and classify such problems into certain types. The current status of existing inventory management and transportation optimization is investigated to study their relationship with production scheduling and analyze the new features generated by the collaborative optimization of production scheduling and transportation and inventory in a supply chain environment, and also analyze the main contradiction and bottleneck problems.

2.3.2 Production Scheduling Coordination Mechanism of Enterprises in a Supply Chain Environment

The collaborative management mechanisms related to production scheduling in a supply chain environment include an information sharing mechanism, and production scheduling management mechanism (including benefit distributing mechanism), which are the basic condition and essential guarantee to realize and implement production scheduling and transportation and inventory collaborative strategies in a supply chain environment. Based on the survey of typical production patterns of the supply chain of DMEs and analysis of the status and new features of the production patterns in a supply chain, the information sharing and production scheduling management mechanisms of the production pattern in a supply chain can be proposed in future research, and the detailed research content is as follows.

1. *The information sharing mechanism of production scheduling in a supply chain environment.* The information sharing mechanism is the basic guarantee of realizing the production scheduling and transportation and inventory collaborative optimization in a supply chain environment. The degree, scope, and mode of information sharing directly impact on the effectiveness of the scheduling plans. The future research mainly studies the key information determination problem (e.g., supply lead time, finished products delivery time) and information sharing mechanism of production scheduling problems in a supply chain environment, and explores the optimal information sharing mechanism of scheduling problems in a supply chain.

2. *The production scheduling management mechanism in a supply chain environment.* The production scheduling in a supply chain environment involves the processing of the inconsistent situations in production scheduling and transportation and inventory among multiple connected nodes in a supply chain. For example, if the operational benefits (e.g., production scheduling optimization, just-in-time, zero inventory) of the client need to be ensured, then the supplier has to assume a higher production cost, transportation cost, or inventory cost, which impacts the overall competitiveness of the supply chain and also weakens the willingness of the supplier and client in the supply chain to cooperate. In view of this, we should focus on the multinode joint production management, joint inventory management, and production and transportation collaborative optimization mechanisms in the supply chain to reduce the operational cost including production, storing, and transportation and optimize production capacity and resource allocation.

2.3.3 Multiobjective Production Scheduling Model Research of DMEs in a Supply Chain

The main difference between the production scheduling problems of DMEs in a supply chain environment and traditional scheduling problems is that the first considers the collaborative optimization of overall scheduling plans of multiple production organizations or functional departments during different stages, and that the optimization objective also involves the entire supply chain, which may contain multiple objectives, such as minimizing production cost, transportation cost, inventory cost, production cycle, shortage stock level, tardiness (maximizing the service responsibility), and so forth. Given the survey of the production manufacturing problems in the supply chain of the DME industry and the literature review, we can propose three types of problems and study the corresponding models.

1. *The research on continuous multiperiod dynamic production scheduling models oriented to the market of core enterprises in a supply chain.* In a multilevel supply chain environment, the multiple nodes of up- and downstream enterprises (or departments) are in a supply–demand relationship. It is essential to study the continuous multiperiod dynamic collaborative optimization production scheduling problems and the multiobjective optimization model within each node in the supply chain, which should be oriented to the market of core enterprises in a supply chain, combining the production plans and production scheduling and lengthening the scheduling cycle. Oriented to the market demand of DME core enterprises that are characterized by multitype, small-batch patterns and flexibility,

the multiobjective mathematical models of continuous multiperiod dynamic production scheduling collaborative optimization (e.g., minimizing production cycle, shortage stock level, tardiness) can be established, and the complexity of the problem needs to be analyzed.

2. *Research on joint collaborative optimization models of production scheduling and inventory and transportation for the entire supply chain.* In a supply chain environment, assuming that the supplier provides the raw materials in batches to save the unit transportation cost, such that the client (i.e., the core enterprise) has a certain amount of material inventory, and also assuming that the finished products of the core enterprise are transported to each storage center in batches, we need to consider comprehensively the factors of raw material distribution, raw material inventory, production scheduling, and finished products delivery in batches and inventory and the multitype, small-batch, and flexibility features of the core enterprise; establish the multiobjective (minimizing production cost, inventory cost, and/or transportation cost, minimizing the production cycle, tardiness, and/or shortage stock level) production scheduling models; and analyze the complexity of the problems.

3. *Research on the new production scheduling problem models specific to a DME supply chain environment.* Some special scheduling problems in DMEs are found in the long-time production optimization research such as the flowshop problem with multiple types, small batches, and inconsistent production processes (different from traditional flowshop problems). We can further study the specific scheduling problem and synthesis problem of transportation and inventory collaborative optimization, establish multiobjective optimization models of these types of problems, and analyze the complexity of the problems.

2.3.4 Research on the Solution Approaches of DME Production Scheduling Problems in a Supply Chain Environment

We need to analyze the characteristics, advantages, disadvantages, and scope of application of the traditional multiobjective solving approaches; combine existing scheduling rules and experience and knowledge in real production to investigate the constructive heuristic algorithms applicable in multiobjective problems; and study heuristic algorithms for multiobjective production scheduling problems in a restricted supply chain based on Pareto optimality. Owing to the main features of multiple types and small-batch production scheduling in DMEs, the scale of the problem is usually large; thus we focus on the multiobjective intelligent algorithms based on the concept of Pareto optimality.

The simulation and assessment of algorithms is not only an essential process in the algorithm analysis, but the algorithms can also be modified in the process of simulation and evaluation. We should design the measurement

indicators that can be used to assess the multiobjective scheduling problem; evaluate the performance of algorithms in the case of a Pareto-optimal
unknown solution set, including the solution quality of the algorithm, diver-
sity (or uniformity of distribution) of the final Pareto approximately optimal
solution set, algorithm convergence, and evolution performance; comprehensively compare each measurement indicator; and construct an evaluation
framework of multiobjective evolutionary algorithms.

2.4 Research Methodology

We can apply mathematical models, traditional multiobjective algorithms,
dispatching rules, heuristics algorithms, multiobjective intelligent algorithms, and case studies to conduct research on the multiobjective production scheduling problems. The detailed research methods and their purposes
are as follows.

2.4.1 Mathematical Modeling Method

Some mathematical modeling methods are proposed as follows:

1. *Using the mathematical modeling method to establish the problem model of
   production scheduling in a supply chain environment.* According to the
   real production environment, the mathematical modeling methods
   (e.g., multiobjective programming, mixed integer programming) are
   utilized to construct mathematical models of the production scheduling problem in a supply chain, in which the real production cases
   are considered to set the constraint conditions, decision variables,
   and intermediate parameters in different types of problem models.
   The cost (e.g., production, transportation, inventory costs), time (e.g.,
   completion time, tardiness), and quality of service (e.g., timely delivery rate, shortage stock level) are set as the objective functions in the
   mathematical models of the problems.

2. *Applying exact algorithms to solve small-size production scheduling problems.* The related exact algorithms applied in linear programming,
   integer programming, mixed integer programming, and dynamic
   programming, such as the relaxed mixed integer programming
   method, branch-and-bound method, column generation method,
   recursive method, and so forth are utilized to find the optimal solutions or lower bound of small-size problems and determine the maximum size of the problem. In addition, the exact solutions within the
   limited scale are used to validate the effectiveness of the methods.

3. *Applying decomposition techniques to decompose complex production scheduling problems.* The decomposition techniques based on machines (resources) (e.g., shifting bottleneck technique) and tasks, or mixed decomposition techniques are utilized to divide the problems into several subproblems solved separately, and the local optimal solutions of these subproblems are integrated by some communication strategy for global optimization. With respect to continuous multiperiod production scheduling problems, the rolling scheduling strategy is used to solve the production scheduling problem in different periods.

4. *Applying evaluation function methods to integrate multiple objectives of the same type.* With respect to problems with the same type of objective functions, for example, the objectives based on cost, the evaluation function method is supposed to be used in solving this type of problem, which can reduce the number of objectives by certain mathematical methods. Some common evaluation function methods include the linear weighted function method, weighted deviation function method, geometric average function method, cost effectiveness function method, and so forth.

5. *Applying other types of traditional multiobjective algorithms to solve multiobjective production scheduling problems.*

6. *For problems with determined preferences of decision makers, the most important or most leading decision objectives are set as the objective functions.* The other objectives are converted to constraint conditions by setting threshold values, so as to decrease the complexity of the problems. Another method is to classify the objective functions into different layers, after satisfying the upper layer of objectives, and then solve the optimal solutions of the next layer of objectives, and so on. For the problems with explicit objective values, the ideal point method is used, in which the decision makers first give an ideal objective value that is smaller than the optimal solution (for minimizing problems), then some norm is introduced into the objective space, and finally the feasible solutions with shortest distance between the ideal point are found in that norm.

2.4.2 Dispatching Rules, Combinatorial Dispatching Rules, and Heuristic Algorithms

Here we propose some dispatching rules and algorithms as follows:

1. *Dispatching rules and combinatorial dispatching rules.* There exist a large number of mature dispatching rules or composite dispatching rules in the traditional production scheduling and combinatorial optimization areas. Based on these existing rules, experiences, and knowledge in a real production environment, we can design

scheduling rules or combinatorial scheduling rules for production scheduling problems in a supply chain environment and analyze the principles of each type of rule and its impact on system performance. This type of method is usually of higher solution efficiency and more targeted, contributes to analyzing the essence and optimization principles of scheduling problems, and also can be used to construct the initial solutions of intelligent algorithms. However, these rules are also subject to some shortcomings, such as low scalability.

2. *Heuristic algorithms.* The production scheduling problems in a supply chain environment are mainly nondeterministic polynomial time (NP)-hard problems. Because the production scheduling problem scale of DMEs is large in general, designing heuristic algorithms is the main trend of solving such problems. Based on various scheduling rules and experiences and knowledge in a real production environment, we can design the heuristic algorithms for production scheduling problems in a supply chain and validate the effectiveness of the heuristics through the known exact solutions in small-size problems, and the results of the heuristics can also be used as the initial solutions of other algorithms (such as multiobjective intelligent algorithms).

2.4.3 Multiobjective Intelligent Algorithms

For the multiobjective programming models of large size DMEs, various effective multiobjective intelligent algorithms can be developed to solve them. The emphasis should be on multiobjective genetic algorithms and multiobjective particle swarm algorithms because these algorithms have been validated to well solve such problems.

1. *Multiobjective genetic algorithms based on Pareto solutions.* Genetic algorithms are a widely used and efficient random searching and optimizing method, which is also a suitable method for solving multiobjective problems. The method of "dominated solutions" is used when maintaining the offspring, that is, when choosing two genes and comparing their fitness: If A is completely better than B, then A dominates B, and B is called the dominated solution; and if A is not dominated by any other solution (vector), then A is called the nondominated solution. The nondominated solutions obtained by the operations of crossover and mutation are the required Pareto-optimal solution set, and the final optimal solution is chosen from this set based on the preferences of decision makers.

2. *Multiobjective particle swarm optimization algorithms.* The particle swarm optimization algorithm is a new evolutionary algorithm

developed in recent years. When solving multiobjective problems, the speed and location of each particle is denoted by a D-dimension vector, and the particles that are no worse than all other particles are the targets of their directions.

2.5 Conclusions

This chapter provides an overview of multiobjective coordinated scheduling problems of discrete manufacturing enterprises in a supply chain environment. Previous research is reviewed in five aspects, including the coordination mechanism of the production system, production plans and production (machine) scheduling, multiobjective production problem-solving techniques, and multiobjective evolutionary algorithm evaluation methods. Based on this review work, it is noticeable that a huge gap exists in the current situation. Moreover, we propose prospective research directions for these problems. We hope this chapter clearly demonstrates the problems and challenges in the DME supply chain and provides new opportunities for researchers to address these challenges. Then, more effective solutions can be developed to deal with practical problems in the industry, thereby strengthening the economic competitiveness of discrete manufacturing enterprises.

Acknowledgments

This research work was supported by the National Natural Science Foundation of China (Nos. 71231004, 71171071, and 71131002). Panos M. Pardalos was partially supported by LATNA laboratory, NRU HSE, RF government grant, ag. 11.G34.31.0057.

References

Agnetis, A., N. G. Hall, and D. Pacciarelli. Supply chain scheduling: Sequence coordination. *Discrete Applied Mathematics*, 154(15): 2044–2063, 2006.
Aliev, R. A., B. Fazlollahi, B. G. Guirimov, and R. R. Aliev. Fuzzy-genetic approach to aggregate production-distribution planning in supply chain management. *Information Sciences*, 177(20): 4241–4255, 2007.

Alvarez, E. Multi-plant production scheduling in SMEs. *Robotics and Computer-Integrated Manufacturing*, 23(6): 608–613, 2007.

Behnamian, J., S. M. T. Fatemi Ghomi, and M. Zandieh. A multi-phase covering Pareto-optimal front method to multi-objective scheduling in a realistic hybrid flowshop using a hybrid metaheuristic. *Expert Systems with Applications*, 36(8): 11057–11069, 2009.

Cardoen, B., E. Demeulemeester, and J. Belien. Optimizing a multiple objective surgical case sequencing problem. *International Journal of Production Economics*, 119(2): 354–366, 2009.

Chang, Y. C. and C. Y. Lee. Machine scheduling with job delivery coordination. *European Journal of Operational Research*, 158(2): 470–487, 2004.

Chern, C. C. and J. S. Hsieh. A heuristic algorithm for master planning that satisfies multiple objectives. *Computers & Operations Research*, 34(11): 3491–3513, 2007.

Chinchuluun, A., P. M. Pardalos, A. Migdalas, and L. Pitsoulis. *Pareto Optimality, Game Theory and Equilibria*. New York: Springer Science+Business Media, 2008.

Dawande, M., H. N. Geismar, N. G. Hall, and C. Sriskandarajah. Supply chain scheduling: Distribution systems. *Production and Operations Management*, 15(2): 243–261, 2006.

Deb, K., A. Pratap, and T. Meyarivan. A fast and elitist multi-objective genetic algorithm: NSGA-II. *IEEE Transactions on Evolutionary Computation*, 6(2): 182–197, 2002.

Demirli, K. and A. D. Yimer. Production-distribution planning with fuzzy costs. In *Proceedings of the Annual Meeting of the North American Fuzzy Information Processing Society*, 2006 (NAFIPS 2006), 702–707, 2006.

Fisher, M. and A. Raman. Reducing the cost of demand uncertainty through accurate response to early sales. *Operations Research*, 44(1): 87–99, 1996.

Frederix, F. An extended enterprise planning methodology for the discrete manufacturing industry. *European Journal of Operational Research*, 129(2): 317–325, 2001.

Hall, N. G. and C. N. Potts. Supply chain scheduling: Batching and delivery. *Operations Research*, 51(4): 566–584, 2003.

Huang, R. H. and C. L. Yang. Solving a multi-objective overlapping flow-shop scheduling. *International Journal of Advanced Manufacturing Technology*, 42: 955–962, 2009.

Iyer, A. V. and M. E. Bergen. Quick response in manufacturer-retailer channels. *Management Science*, 43(4): 559–570, 1997.

Jayaraman, V. and H. Pirkul. Planning and coordination of production and distribution facilities for multiple commodities. *European Journal of Operational Research*, 133(2): 394–408, 2001.

Kaminsky, P. and O. Kaya. Inventory positioning, scheduling and lead-time quotation in supply chains. *International Journal of Production Economics*, 114(1): 276–293, 2008.

Kanyalkar, A. P. and G. K. Adil. An integrated aggregate and detailed planning in a multi-site production environment using linear programming. *International Journal of Production Research*, 43(20): 4431–4454, 2005.

Knowles, J. D. *Local-Search and Hybrid Evolutionary Algorithms for Pareto Optimization*. PhD thesis. Department of Computer Science, University of Reading, UK. January, 2002.

Lee, C. Y. and Z. L. Chen. Machine scheduling with transportation considerations. *Journal of Scheduling*, 4(1): 3–24, 2001.

Leng, M. and M. Parlar. Lead-time reduction in a two-level supply chain: Non-cooperative equilibria vs. coordination with a profit-sharing contract. *International Journal of Production Economics*, 118(2): 521–544, 2009.

Li, C. L., G. Vairaktarakis, and C. Y. Lee. Machine scheduling with deliveries to multiple customer locations. *European Journal of Operational Research*, 164(1): 39–51, 2005.

Liang, T. F. Applying fuzzy goal programming to production/transportation planning decisions in a supply chain. *International Journal of Systems Science*, 38(4): 293–304, 2007.

Low, C., Y. Yip, and T. H. Wu. Modeling and heuristics of FMS scheduling with multiple objectives. *Computers & Operations Research*, 33(3): 674–694, 2006.

Mazdeh, M. M., S. Shashaani, A. Ashouri, and K. S. Hindi. Single-machine batch scheduling minimizing weighted flow times and delivery costs. *Applied Mathematical Modeling*, 35(1): 563–570, 2011.

Mula, J., D. Peidro, M. Díaz-Madroñero, and E. Vicens. Mathematical programming models for supply chain production and transport planning. *European Journal of Operational Research*, 204(3): 377–390, 2010.

Pasupathy, T., C. Rajendran, and R. K. Suresh. A multi-objective genetic algorithm for scheduling in flow shops to minimize the makespan and total flow time of jobs. *International Journal of Advanced Manufacturing Technology*, 27: 804–815, 2006.

Pei, J., X. Liu, P. M. Pardalos, W. Fan, L. Wang, and S. Yang. Solving a supply chain scheduling problem with non-identical job sizes and release times by applying a novel effective heuristic algorithm. *International Journal of Systems Science*, 2014a. doi: 10.1080/00207721.2014.902553.

Pei, J., X. Liu, P. M. Pardalos, W. Fan, S. Yang, and L. Wang. Application of an effective modified gravitational search algorithm for the coordinated scheduling problem in a two-stage supply chain. *The International Journal of Advanced Manufacturing Technology*, 70(1–4): 335–348, 2014b.

Rizk, N., A. Martel, and S. D'Amours. Multi-item dynamic production-distribution planning in process industries with divergent finishing stages. *Computers & Operations Research*, 33(12): 3600–3623, 2006.

Rowe, J., K. Jewers, J. Sivayogan, A. Codd, and A. Alcock. Intelligent retail logistics scheduling. *AI Magazine*, 17(4): 31–40, 1996.

Sahin, F., E. P. Robinson, and L. L. Gao. Master production scheduling policy and rolling schedules in a two-stage make-to-order supply chain. *International Journal of Production Economics*, 115(2): 528–541, 2008.

Sawik, T. Coordinated supply chain scheduling. *International Journal of Production Economics*, 120(2): 437–451, 2009.

Selim, H., C. Araz, and I. Ozkarahan. Collaborative production-distribution planning in supply chain: A fuzzy goal programming approach. *Transportation Research Part E: Logistics and Transportation Review*, 44(3): 396–419, 2008.

Selvarajah, E. and G. Steiner. Batch scheduling in a two-level supply chain—A focus on the supplier. *European Journal of Operational Research*, 173(1): 226–240, 2006.

Tavakkoli-Moghaddam, R., A. Rahimi-Vahed, and A. H. Mirzaei. A hybrid multi-objective immune algorithm for a flow shop scheduling problem with bi-objectives: Weighted mean completion time and weighted mean tardiness. *Information Sciences*, 177(22): 5072–5090, 2007.

Torabi, S. A. and E. Hassini. An interactive possibilistic programming approach for multiple objective supply chain master planning. *Fuzzy Sets and Systems*, 159(2): 193–214, 2008.

Xia, W. and Z. Wu. An effective hybrid optimization approach for multi-objective flexible jobshop scheduling problems. *Computers & Industrial Engineering*, 48(2): 409–425, 2005.

Xie, Y., D. Petrovic, and K. Burnham. A heuristic procedure for the two-level control of serial supply chains under fuzzy customer demand. *International Journal of Production Economics*, 102(1): 37–50, 2006.

Yagmahan, B. and M. M. Yenisey. Ant colony optimization for multi-objective flow shop scheduling problem. *Computers & Industrial Engineering*, 54(3): 411–420, 2008.

Zegordi, S. H., I. N. Kamal Abadi, and M. A. Beheshti Nia. A novel genetic algorithm for solving production and transportation scheduling in a two-stage supply chain. *Computers & Industrial Engineering*, 58(3): 373–381, 2010.

Zhang, G., X. Shao, P. Li, and L. Gao. An effective hybrid particle swarm optimization algorithm for multi-objective flexible job-shop scheduling problem. *Computers & Industrial Engineering*, 56(4): 1309–1318, 2009.

Section II

Supply Chain Network Optimization

3

Integrated Production/Distribution/ Routing Planning for Supply Chain Networks: A Review

Lei Lei, Rosa Oppenheim, Lian Qi, Hui Dong,
Kangbok Lee, and Shengbin Wang

CONTENTS

ABSTRACT This chapter presents an overview of existing solution methodologies for integrated operations planning problems in supply chain networks involving production, inventory, distribution, and routing. We take into account problems dealing with operational decisions and classify them according to their characteristics, such as time constraints and routing decisions. Various methodologies are presented and their possible integrations and combinations are discussed. Finally, future research directions are proposed.

KEY WORDS: *distribution, integrated operations planning, inventory, production, routing, supply chain.*

3.1 Introduction

A supply chain is defined as an integrated business process with bidirectional flows of products, information, cash, and services, between tiers of suppliers, manufacturers, logistics partners, distributors, retailers, and customers. Because of fast changes in the marketplace and the rapid expansion of supply chains (Eksioglu et al., 2007), ensuring highly coordinated production, inventory, and distribution over a multi-echelon supply chain network is vital, and has an immediate impact on customer services and profit margins. This importance will continue to increase along with the following trends.

Globalization. All functions in a supply chain network, such as procurement, production, distribution, and consumption, have now become more globalized. Most multinational firms have business facilities located over multiple continents, with many local markets to serve; face the need for emerging market penetration and the challenge of capacity shortages and rising shipping costs; and are constantly confronting environmental/sustainability concerns. At the same time, the promises and flexibility of third-party logistics and subcontracting opportunities offer a great incentive to expand supply chains globally. As supply chains expand, the need to ensure a more precise match between demand and supply increases the importance of integrated operations planning.

Pressure on lead time reduction and profit margin improvement. Because customer demand for both products and services typically changes over time, time-to-market is more important than ever to meet the expectations of demanding customers. For most supply chains, production is not the only major process to be considered; there are many other stages, such as sourcing, distribution, inventory, packaging, and order processing that together could account for a significant portion of the lead time. A less coordinated supply chain process could easily diminish or eliminate the profit margin and lead to poor customer service.

Advances in information technology. Advances in information technology during the past two decades have significantly improved data visibility (e.g., inventory visibility and shipping status) and information accessibility along the supply chain. Data can be automatically collected, retrieved, and manipulated in various ways and shared by many supply chain partners (e.g., through radio frequency identification [RFID]). Furthermore, today's computing power allows us to solve, relatively easily and more rapidly, larger scale integrated operations planning problems that were difficult, if not impossible, only a few decades ago when optimization problems of a combinatorial nature were considered computationally intractable.

Serving the needs of emerging noncommercial supply chains. A network for disaster relief operations is a typical illustration of a noncommercial supply chain. Disaster relief and emergency logistics (e.g., in response to Hurricane Katrina in Louisiana in 2005, the tsunami in Japan in 2011, and Hurricane Sandy in New Jersey and New York in 2012) usually cannot be handled effectively by a single state or a single local government. Today's Internet allows the need for disaster relief to be communicated cross-country and internationally within minutes of an event and the rapid formation of disaster relief supply chains for quick response to people in the affected areas. A highly effective and fully integrated production and distribution operation that pulls supplies from different industries and states to ensure delivery of these resources to the people in an affected area is critical to human well-being.

In this chapter we focus on the solution methodologies for solving various *integrated/coordinated* production and distribution operations planning problems reported in the current literature. This survey does not focus on results related to decisions for supply chain designs (e.g., facility location and/or facility capacity), or on those results that deal only with a single operation such as inventory, or routing, or production scheduling, but rather addresses issues unique to process integration.

There have been several survey papers dealing with integrated operations problems, each with its own focus. Among these, the pioneer review by Thomas and Griffin (1996) defines a generic structure for a supply chain network and classifies published results at both the strategic planning level and the operational planning level, the latter of which falls into our scope. The models related to operational planning are classified into buyer and vendor coordination, production–distribution coordination, and inventory–distribution coordination; up through the time of this study, most researchers, because of limitations on computational capability, have decomposed such multistage problems into several two-stage problems that are then solved separately. Erenguc et al. (1999) review the studies on managing supply chain networks with three distinct stages consisting of suppliers, plants, and distribution centers and focus on the results for joint operational decision making across the three stages. Decisions that need to be jointly made regarding optimizing production–distribution planning are discussed. Sarmiento and Nagi (1999) consider integrated production–distribution planning systems at both the strategic and tactical levels with an explicit consideration of transportation. They classify the problems based on the type of decisions being modeled (e.g., decisions on production, distribution, or inventory management) and on the number of locations per echelon in the model. Three categories of two-echelon models are identified, and the differences between such models and those in classical Inventory Routing studies are discussed. Fahimnia et al. (2008b) review existing production–distribution planning models and provide a table summarizing 19 papers according to problem attributes (e.g., numbers of plants, distribution centers, and customers, multiperiods, multiproducts, routing), types of modeling approaches (e.g., mathematical programming, optimization, simulation, and combinations of these), and the solution methods applied.

There are also two recent survey papers on integrated operations planning: Mula et al. (2010) and Fahimnia et al. (2013). Mula et al. (2010) cite 44 papers published since 1985 among the 54 references and classify these works based on the decision levels (e.g., strategic, tactical, and operational), modeling approach (e.g., linear programming and multiobjective integer linear programming), objective (e.g., total cost and customer satisfaction), level of information sharing (e.g., production cost, lead time, inventory level, and demand), and solution methodologies. Fahimnia et al. (2013) cite 139 papers related to integrated operations planning and classify these papers by two criteria: complexity of the network structure and solution methodologies. Interestingly, in spite of the large number of references listed in these surveys, only 19 papers were common to both surveys. However, there is no analysis in either survey on the relationship between problem structures and the methodologies reported in these works.

Unlike the existing surveys, we focus here on the relationships between the problem structures and solution methodologies. Such a survey provides information to researchers on the solution approaches, developed for solving problems defined over different types of network structures, and their effectiveness. We classify the integrated operations planning problems into four categories. For each category, we present a basic mathematical model and, based on the properties of the respective network structure, analyze the existing solution methodologies. To define these categories, two attributes are used: *time constraint* and *routing*. Most integrated operations planning problems involve multiple time periods. For each period, the ending inventory level, production quantity, and distribution amount must be determined. Because a continuous time scale within a period has to be considered in some studies to describe time constraints such as arbitrary delivery deadlines or travel times, there is a need to model the time constraints explicitly. Note that without such explicit modeling of time constraints, as many studies in the past have done, we often have to assume that any quantity produced in one period is delivered to customers in the same period, which leads to a gap between the models and the real-world practice. For those studies involving direct shipment between suppliers and customers, we allow the shipping

TABLE 3.1

Categories of the Integrated Operations Planning Problems

Issues in the Literature Problem Categories	Production Issues	Distribution Issues	Time Constraints	Routing Issues
Production and distribution problem (PDP)	X	X		
PDP with time constraints (PDPT)	X	X	X	
PDP with routing (PDPR)	X	X		X
PDP with routing and time constraints (PDPRT)	X	X	X	X

capacity to be defined as either the maximum outgoing flow amount or the fleet size and/or capacity of vehicles. For the studies in which one vehicle may visit several customers in one trip, we allow vehicle routing issues to be explicitly included in the model. We categorize the problems into four categories in Table 3.1.

We also refer readers to another survey by Yossiri et al. (2012), in which the authors categorize the studies according to their inclusion of decision variables related to the flow quantity of production, inventory, distribution, and routing.

The remainder of this chapter is organized as follows. In Section 3.2, we introduce the basic assumptions of the integrated operations planning problems. In particular, the assumptions of each of the four categories shown in Table 3.1—PDP, PDPT, PDPR, and PDPRT—are presented. In Section 3.3, we focus on the studies and solution approaches for the integrated production and distribution problems, PDPs, that involve no routing and time constraints; most of the papers from the related literature belong to this class of problems. In Section 3.4, we extend PDP to include time constraints, and in Section 3.5 we extend PDP to include routing issues. In Section 3.6, we review those existing studies that include both time constraints and routing, an area where the design of effective solution methodologies is much more challenging. Discussion and future research directions are presented in Section 3.7.

3.2 Assumptions and Preliminaries

In this section, we introduce the common assumptions and notation used to define the four categories of problems: PDP, PDPT, PDPR, and PDPRT. For each assumption, we then discuss its extensions or variations that are found in the literature.

Product and time dimension. We consider the multiproduct problem (i.e., with multiple commodities) over a given planning horizon of multiple time periods.

Network structure and material flow. The supply chain network has three stages: manufacturers, distribution centers (DCs), and customers, as shown in Figure 3.1. Each customer has a certain demand to be fulfilled in each period. Both manufacturers and DCs hold inventories of products. Manufacturers produce and fill their own inventories and send products to DCs, which in turn send the products to customers.

Extensions or variations in the literature. There exist suppliers to provide manufacturers with raw material.

There exist third parties that serve as contract manufacturers or DCs. The third parties usually charge higher prices than regular players.

In some cases, manufacturers may deliver the product directly to customers.

Production and transportation capacity. Each manufacturer has a maximum production capacity (i.e., the maximum quantity that it is able to produce) in each period. Both manufacturers and DCs have a maximum transportation capacity (i.e., the maximum outgoing flow quantity) in each period.

Extensions or variations in the literature. A manufacturer's production capacity can be increased at an additional fixed and/or variable cost (e.g., overtime work).

Transportation capacity can be defined by the vehicle attributes (e.g., the fleet size, the vehicle loading capacity, the maximum number of trips, the total working hours in one period, etc.).

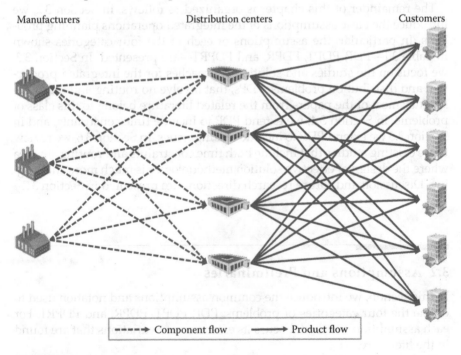

Manufacturers   Distribution centers   Customers

----▶ Component flow   ——▶ Product flow

FIGURE 3.1
Network structure and material flow.

Customer demand fulfillment and on-time delivery. All customer orders must be fulfilled on time, and no customer carries inventory.

Extensions or variations in the literature. If an order is not fulfilled on time, it is lost (called a *lost-sale*).

If an order is not fulfilled on time, it can be fulfilled later with a penalty cost (either as a *backorder* delivered in a subsequent period, or as a late shipment within the same period).

Cost components. Each manufacturer has a fixed, and variable, cost of production, and each DC has a fixed, and variable, cost for handling the product. Both manufacturers and DCs incur inventory holding costs. The shipments from manufacturers to DCs, and from DCs to customers, result in a shipping cost.

Extensions or variations in the literature.

- When raw materials are required, the purchase cost is considered.
- When a third party is involved, the respective costs (e.g., contract fees) are included.
- If a late delivery (backorder) is allowed, the relevant penalty cost is included.
- If a lost-sale is allowed, the shortage penalty is included.

Although a representative mathematical model for each of the following sections is built upon these basic assumptions, its variations are introduced as we discuss individual papers.

Throughout this survey, we use the following notation: let $M = \{m\}$, $B = \{i\}$, $J = \{j\}$, and $K = \{k\}$ denote the set of manufacturing facilities, the set of distribution/transshipment centers (DCs), the set of customers, and the set of products ordered by customers, respectively. When routing decisions are involved, let $V(m)$ denote the set of vehicles of manufacturer m. Let $T = \{t\}$ denote the set of periods. For simplicity, $\forall m$, $\forall i$, $\forall j$, $\forall k$, $\forall v$, and $\forall t$ may be used instead of $\forall m \in M$, $\forall i \in B$, $\forall j \in J$, $\forall k \in K$, $\forall v \in V(m)$, and $\forall t \in T$.

3.3 The Production and Distribution Problem

The production and distribution problem (PDP) is primarily concerned with coordinating production and outbound distributions to minimize the total costs associated with production, inventory, and transportation over a discrete multiperiod planning horizon. Because PDP does not explicitly include the routing and shipping times, the models for PDP involve only inventory flow balance, facility capacity, and transportation capacity constraints (e.g., Thomas and Griffin, 1996).

To define the mathematical model for the PDP formally, we introduce the following notation: for any given period t, let $C^k_{m,t}$ be the production capacity of manufacturer m for product k, $C_{a,b,t}$ be the transportation capacity from location a to location b for $(a, b) \in M \times B \cup B \times J$, and $d^k_{j,t}$ be the demand for product k by customer j. Let $I^k_{a,0}$ be the initial inventory of product k at location a for $a \in M \cup B \cup J$. For decision variables, let $W_{a,b,t}$ and $Z^k_{m,t}$, respectively, be the binary variables denoting the decision for a flow from location a to location b for $(a, b) \in M \times B \cup B \times J$ in period t, and the decision for a production batch for product k by manufacturer m in period t. Let S, Q, P, and I, each with proper superscript and subscript indices, be continuous variables denoting the shortage amount, flow quantity, production quantity, and inventory level, respectively. For example, $Q^k_{m,i,t}$ denotes the flow quantity of product k from manufacturer m to DC i in period t. In addition, we use $M||J$, and $B||J$, to denote a network involving only manufacturers and customers, and distribution centers and customers, respectively, and $M||B||J$ to denote a network involving all three stages. A basic PDP model can then be described as follows:

Minimize: $G\left(W_{m,i,t}, W_{i,j,t}, Z^k_{m,t}, S^k_{j,t}, Q^k_{m,i,t}, Q^k_{i,j,t}, P^k_{m,t}, I^k_{m,t}, I^k_{i,t}, I^k_{j,t}\right)$ 　　　(3.1)

s.t.

$$I^k_{m,t-1} + P^k_{m,t} - \sum_{\forall i} Q^k_{m,i,t} = I^k_{m,t}, \quad \forall m, k, t \tag{3.2}$$

$$I^k_{i,t-1} + \sum_{\forall m} Q^k_{m,i,t} - \sum_{\forall j} Q^k_{i,j,t} = I^k_{i,t}, \quad \forall i, k, t \tag{3.3}$$

$$I^k_{j,t-1} + \sum_{\forall i} Q^k_{i,j,t} - \left(d^k_{j,t} - S^k_{j,t}\right) = I^k_{j,t}, \quad \forall j, k, t \tag{3.4}$$

$$P^k_{m,t} \le C^k_{m,t} \cdot Z^k_{m,t}, \quad \forall m, k, t \tag{3.5}$$

$$\sum_{\forall k} Q^k_{m,i,t} \le C_{m,i,t} \cdot W_{m,i,t}, \quad \forall m, i, t \tag{3.6}$$

$$\sum_{\forall k} Q^k_{i,j,t} \le C_{i,j,t} \cdot W_{i,j,t}, \quad \forall i, j, t \tag{3.7}$$

$W_{m,i,t}, W_{i,j,t}, Z^k_{m,t} \in \{0,1\}, \quad S^k_{j,t}, Q^k_{m,i,t}, Q^k_{i,j,t}, P^k_{m,t}, I^k_{m,t}, I^k_{i,t}, I^k_{j,t} \ge 0 \quad \forall m, i, j, k, t$ (3.8)

The objective function (Equation 3.1) minimizes the total operations cost, consisting of raw materials, facility setup, production, inventory, and transportation costs. The constraints in Equations 3.2 through 3.4 ensure the flow balances at the manufacturing facilities, DCs, and customer sites, respectively, while the constraints in Equations 3.5 through 3.7 are network capacity constraints.

Although special cases of PDP, such as the classical transportation problem and the transshipment problem, can be solved in strongly polynomial time, the general version of the PDP is difficult to solve. More precisely, the multiproduct PDP defined by Equations 3.1 through 3.8 is strongly nondeterministic polynomial time (NP)-hard, because a special case of this PDP is a multiproduct multiperiod lot-sizing problem that has been proved to be strongly NP-hard by Chen and Thizy (1990). Therefore, a general version of PDP could require an excessive amount of computational time to verify the solution optimality when the network size becomes large.

In this section, we focus on the existing solution methodologies for variations of the PDP defined by Equations 3.1 through 3.8, and classify them into three categories. The first one is heuristic and metaheuristic algorithms, in which a solution (or a set of solutions) is constructed by relatively simple rules and then improved through an iterative process. The other two are both mathematical programming-based solution approaches, and differ on how the PDP model is relaxed: constraints relaxation approaches and variables relaxation approaches. Note that although the routing decision is not considered in this section, we do include those problems that assume fixed routing.

3.3.1 Heuristic and Metaheuristic Algorithms

Because of the intractability of the general PDP, feasible solutions with acceptable quality and minimal solution time have been commonly discussed in the literature. Representative solution approaches in this category are greedy heuristics and genetic algorithms.

Park (2005) proposes a two-phase heuristic for solving a multiproduct PDP defined on an $M||J$ network to maximize the total profit. The phase I problem is formed by aggregating the demand of all customers in each period, defined by $D_t^k = \sum_{\forall j} d_{j,t}^k$ and then replacing constraint (Equation 3.4) by $I_{t-1}^k + Q_t^k - D_t^k = I_t^k$, $\forall k$, t, in the model, which reduces the problem to a single-customer multi-period model and allows one to quickly determine the values of $P_{m,t}^k$ by solving a production lot-sizing problem (Fumero and Vercellis, 1999) with constant production capacity. All unsatisfied demand is penalized as shortage and no backorder is considered. Given $P_{m,t}^k$, the author then solves a distribution problem in phase II to determine the values of $Q_{m,j,t}^k$ by applying a bin-packing heuristic together with local improvement procedures that consolidate partial loads by shifting shipping periods and reducing the level of stock-out using leftover production capacity. Through computational experiments on 21 test problems of three sizes, this heuristic achieves an error gap, or a difference between the optimal and heuristic solutions, of 5.6%–6.8% for small-size cases and no more than 9.2% for all the test cases. The computation time is less than 3 seconds for small cases and no more than 1200 seconds for large cases.

Ahuja et al. (2007) study a two-echelon $M||J$ single product PDP with a single sourcing constraint, which means that each customer receives shipment from at most one supplier in each period. In addition to constraints in Equations 3.2 through 3.7, the authors also include a constraint on inventory perishability, so that the maximum inventory time for the product is bounded by a given constant N. Thus, at any period t, the ending inventory at DC i, $i \in I$, cannot exceed its future demand from all customers in the next N periods, or $I_{i,t} \leq \sum_{n=1}^{N} \sum_{\forall j} Q_{i,j,t+n}$. The resulting PDP is decomposed into two subproblems. One includes only binary facility–customer assignment variables, and the other includes variables for transportation flow and inventory levels. A proposed greedy heuristic is used to assign the facility–customer pairs, on which a very-large-scale-neighborhood (VLSN) search heuristic is applied to improve the quality of the solution. Extensive tests on randomly generated problem sets are conducted, and the error gap obtained by comparing the heuristic with the best lower bound obtained by CPLEX within 15 minutes of central processing unit (CPU) time is less than 3% in all cases. The authors also report that their error gaps have a decreasing tendency as the number of customers is increased, and it is less than 0.1% in the largest size case. The computation time is less than 40 seconds in all cases.

Some researchers consider PDP with extensions such as fixed routes for transportation or direct shipment. Lei et al. (2006) investigate an integrated production, inventory, and distribution routing problem encountered from the practices of after-merge operations of a chemical company. A two-phase approach is proposed, in which the phase I problem is defined by assuming direct shipment between manufacturing plants and customers. The assumptions on direct shipments allow the authors to solve an optimization problem with a significantly reduced complexity, which yields a feasible solution to the original problem. The problem in phase II is to improve the solution from phase I and is modeled as a shortest path problem on a directed acyclic graph. An empirical study that evaluates the computational performance of this solution approach is also reported. Liu et al. (2008) study a multiproduct packing and delivery problem with a single capacitated truck and a fixed sequence of customer locations. The authors first apply a network flow-based polynomial time algorithm to solve the problem assuming no split deliveries and then allow the split delivery to improve the truck efficiency by using a greedy heuristic with a time complexity of $O(|J|^3 \log |J|)$. In both papers, optimal solutions of the special cases (with restriction) are modified to obtain feasible solutions to the original problems.

During the past two decades, the genetic algorithm (GA), inspired by the process of natural evolution, has been quickly gaining in popularity. In Jang et al. (2002), the problem of production and distribution planning over a three-echelon $M||B||J$ network is considered. Constraints similar to those in Equations 3.1 through 3.7 are included and a material transform factor Γ is used to define the rate of raw materials consumption: $I_{m,t-1} + P_{m,t} - \sum_{\forall i} \Gamma_{mi} \cdot Q_{m,i,t} = I_{m,t}, \forall m, t$. The solution of the proposed GA algorithm is compared with that obtained by CPLEX. Among randomly generated test problems, the solution time of GA is quite stable, averaging from 334 to 546 seconds, while that required by the CPLEX solver exhibits exponential growth with respect to problem size, from 32 to 67,854 seconds to obtain the optimal solutions. The proposed GA also demonstrates strong performance, with an average error gap of 0.2%. Gen and Syarif (2005) propose a GA-based approach for their $M||J$ network. A new solution approach called the spanning tree-based genetic algorithm is presented together with the fuzzy logic controller concept for auto-tuning the GA parameters. The proposed method is also compared with a traditional spanning tree-based approach. This comparison shows that the proposed approach achieves a better result in every experiment, with an average improvement from 0.05% to 0.65% for six different settings. Kannan et al. (2010) develop an $M||B||J$ network model for battery recycling. Besides production, inventory, and transportation costs, the objective function includes additional cost factors for recycling such as collection, disposal, and reclaiming costs. The authors introduce a heuristic-based genetic algorithm to solve the problem and compare the result with that obtained by the General Algebraic Modeling System (GAMS), a commercial solver. In experiments with different problem sizes and heuristic parameters (population and iteration), the maximum error observed is 7.4% compared with the results from GAMS. Moreover, the average computation time of the GA-based approach is less than 315 seconds for the largest problem whereas that by GAMS is at least 2800 seconds for the smallest problem.

3.3.2 Constraints Relaxation-Based Approaches

Another popular solution approach to PDP in the current literature is to relax a subset of constraints to make the relaxed problem easier to solve. The major approach in this regard is the well-known Lagrangean relaxation, by which difficult constraints are placed into the objective function with coefficients called Lagrangean multipliers so that the resulting problem is "easily solvable." One example of such an easily solvable problem is a network flow problem (Ahuja et al., 1993). Another important approach is based on problem decomposition, by which a subset of constraints is temporarily simplified or removed from the original model to make the remaining problem decomposable. When a Lagrangean relaxation is adapted to achieve the decomposition, the resulting process is called Lagrangean decomposition. In constraints relaxation-based approaches, identifying the constraints to be relaxed and ensuring that the search converges to the optimal or near-optimal solution quickly are two critical steps for achieving the quality and effectiveness of such solution approaches. For example, in the basic model defined by Equations 3.1 through 3.8, when we relax the constraint in Equation 3.3 and incorporate it in the objective function with penalty factors, the problem is decomposed into two problems as follows:

Minimize: $G^1\left(W_{m,i,t}, Z^k_{m,t}, Q^k_{m,i,t}, P^k_{m,t}, I^k_{m,t}\right)$ s.t.   (Equations 3.2, 3.5, and 3.6)

Minimize: $G^2\left(W_{i,j,t}, S^k_{j,t}, Q^k_{i,j,t}, I^k_{i,t}, I^k_{j,t}\right)$ s.t.   (Equations 3.4, 3.7, and 3.8)

where both G^1 and G^2 include the penalty terms for violating constraint (3.3).

Yung et al. (2006) use constraints relaxation to solve a multiproduct single-period PDP, and thus the time index t is dropped from all the notations, defined upon an $M||J$ network. Their study involves decisions on production and transportation, as well as on lot-sizing and order quantity. The average inventory level is used to define the inventory cost, and variables z^k_m and x^k_{mj} are added to denote production lot size and shipping quantity for product k. The model contains flow balance constraints similar to those in Equations 3.2 through 3.4, and capacity constraints similar to those in Equations 3.5 through 3.7. However, the objective function includes terms P^k_m/z^k_m as the number of production lots for product k at manufacturer m and terms Q^k_{mj}/x^k_{mj} as the number of shipments of product k from m to j, which lead to a nonlinear objective function that is neither convex nor concave. To apply Lagrangean relaxation, an artificial variable R_{mj} is utilized, where:

$$\sum_k Q^k_{mj} = R_{mj} \qquad (3.9)$$

and redundant constraints $\sum_k P^k_m = \sum_j R_{mj}$, $\sum_k d^k_j = \sum_m R_{mj}$, and $0 \le R_{mj} \le \sum_k d^k_j$ are added to the model. By relaxing the constraint in Equation 3.9, the original model is decomposed into two independent submodels. The first one deals with joint decisions on production and lot-sizing and thus contains variables P^k_m, z^k_m, and the aggregated transportation flow, R_{mj}. In the second model, the constraints for transportation planning involving Q^k_{mj} and ordering quantity x^k_{mj} are included. By continuously updating the Lagrangean multipliers and the artificial variables, two subproblems are iteratively solved. The test result is compared with that obtained by F_{mincon}, a nonlinear programming tool box in MATLAB® 6.1. Among seven problem settings, F_{mincon} cannot terminate for three cases whereas the proposed algorithm is able to solve all of the cases. In terms of the solution performance, the proposed algorithm saves 1.5%–8% in cost, with less CPU time, over what F_{mincon} achieves for all the cases solved.

Eksioglu et al. (2007) consider a variation of multiproduct multiperiod PDP on an $M||J$ network where only the production facility carries an inventory and there are no capacity limits for inventory and transportation. The model contains flow balance constraints:

$$I_{m,t-1}^k + P_{m,t}^k - \sum_{\forall j} Q_{m,j,t}^k = I_{m,t}^k \qquad (3.10)$$

instead of Equations 3.1 and 3.2. Because the model does not allow shortages, it has

$$\sum_{\forall m} Q_{m,j,t}^k = d_{j,t}^k \qquad (3.11)$$

instead of Equation 3.4, and capacity constraint in Equation 3.5 with binary indicator variables for production. Unlike the previous solution approach, which uses redundant aggregated variables, this approach introduces redundant disaggregated variables. The authors reformulate the original model by introducing a new variable, $Q_{mjt\tau}^k$, which defines the amount of product k from manufacturer m to customer j to satisfy demand in period τ using the quantity produced in period t, where $t \leq \tau$. Thus, the original variables can be expressed by new variables as follows:

$$P_{mt}^k = \sum_{j=1}^{J} \sum_{\tau=t}^{T} Q_{mjt\tau}^k, \quad \forall m, k, t \qquad (3.12)$$

$$Q_{mjt}^k = \sum_{s=1}^{t} Q_{mjst}^k, \quad \forall m, j, k, t \qquad (3.13)$$

$$I_{mt}^k = \sum_{j=1}^{J} \sum_{s=1}^{t} \sum_{\tau=t+1}^{T} Q_{mjs\tau}^k, \quad \forall m, k, t \qquad (3.14)$$

By using the constraints in Equations 3.12 through 3.14, the original model becomes a facility location problem. The authors then show that the linear programming (LP) relaxation of the location model provides a tighter lower bound than the LP relaxation of the original model. Lagrangean decomposition is applied to the resulting location problem by introducing $z_{mjt\tau}^k$, clone or copy of $Q_{mjt\tau}^k$:

$$Q_{mjt\tau}^k = z_{mjt\tau}^k \qquad (3.15)$$

Accordingly, redundant constraints for $z^k_{mjt\tau}$:

$$\sum_{m=1}^{M}\sum_{t=1}^{\tau} z^k_{mjt\tau} = d^k_{j\tau} \tag{3.16}$$

$$\sum_{j=1}^{J}\sum_{k=1}^{K}\sum_{\tau=1}^{T} z^k_{mjt\tau} \leq C_{mt} \tag{3.17}$$

$$z^k_{mjt\tau} \geq 0 \tag{3.18}$$

are then added into the model. By relaxing Equation 3.15 using a Lagrangean multiplier, the model is decomposed into two subproblems. The first one containing $Q^k_{mjt\tau}$ is an uncapacitated multiproduct problem and is further decomposed into $|K|$ single product subsubproblems that are NP-hard but solvable by dynamic programming. On the other hand, the second one containing $z^k_{mjt\tau}$ can be modeled as an LP problem. For test problems of large sizes, the subproblems are solved by using the primal-dual algorithm and the total running times vary from 4 to 87 CPU seconds with empirical error gaps no more than 5%.

Karakitsiou and Migdalas (2008) consider a single product PDP defined on an $M||J$ network. The model has flow balance constraints similar to Equations 3.2 through 3.4, and capacity constraints similar to Equations 3.5 through 3.7. Defining a new variable:

$$r_{m,t} = \sum_{j} Q_{m,j,t} \tag{3.19}$$

the inventory flow balance constraint at m is replaced by

$$I_{m,t-1} + P_{m,t} - r_{m,t} = I_{i,t} \tag{3.20}$$

and the transportation capacity constraint is replaced by

$$0 \leq r_{m,t} \leq C^S_{m,t} \tag{3.21}$$

where $C^S_{m,t}$ is the maximum outbound shipping quantity. Moreover, a redundant constraint

$$\sum_{m} r_{m,t} = \sum_{j} d_{j,t} \tag{3.22}$$

is added. To apply Lagrangean decomposition, a clone variable of $r_{m,t}$, denoted as $z_{m,t}$, is introduced:

$$r_{m,t} = z_{m,t} \tag{3.23}$$

so that the constraint in Equation 3.20 can be replaced by

$$I_{m,t-1} + P_{m,t} - z_{m,t} = I_{i,t} \tag{3.24}$$

$$0 \le z_{m,t} \le C^S_{m,t} \tag{3.25}$$

By relaxing Equation 3.23 and using Lagrangean multipliers, the original model is decomposed into two independent parts. The first one deals with variables $P_{m,t}$, $I_{i,t}$, and $z_{m,t}$ together with the constraints in Equations 3.5, 3.24, and 3.25, and the second one deals with $Q_{m,j,t}$ and $r_{m,t}$ together with the constraints in Equations 3.4, 3.19, 3.21, and 3.22. The first subproblem can be further decomposed, over the manufacturing facilities, into $|M|$ sub-subproblems that can each be modeled as a linear programming problem. The second subproblem can also be further decomposed, over the time horizon, into $|T|$ sub-subproblems, each as a network flow problem. To check the quality of the solutions produced by the Lagrangean relaxation, the results are compared with the optimal solution obtained by the GNU Programming Kit (GLPK) solver, a free and open source software. For six randomly generated problems involving 30–1200 nodes, the empirical error gaps are no more than 6% and the required computation time is no more than 350 seconds.

3.3.3 Variables Relaxation-Based Approaches

During the past decade, the variables relaxation-based approaches, in which a selected subset of integer variables is relaxed so that the reduced problem can be relatively easy to solve, have gained a significant amount of attention from researchers. While the Lagrangean relaxation procedures aim at reducing the duality gaps, most variables relaxation-based approaches focus on reducing the suboptimality due to rounding linear values to integers.

Dogan and Goetschalckx (1999) introduce a multiproduct multiperiod PDP model involving strategic decisions on the network and detailed production planning on the machine level along with deterministic seasonal customer demands. The network under consideration includes candidates for suppliers, potential manufacturing facilities, and DCs with multiple possible configurations and customers. The manufacturing facilities have alternative facility types, which introduce binary variables for the facility selections, and integer variables are used to define the number of machines used in each facility during each period. In addition to the ending inventory, the authors also consider the work-in-process inventory that defines part of the inventory holding cost. Replenishment of raw material may happen more than once during each period. Transportation flow quantities and production quantities on each machine at each facility are also decision variables. Benders decomposition is used as the solution methodology. In the mixed integer master problem, the status of the facilities, the production lines, and the production and inventory quantities are determined. The reduced problem becomes a minimum-cost transportation flow problem, and its optimal cost is added to the mixed integer master problem to find a feasible schedule satisfying the obtained flow cost. The search terminates when the master problem can find no lower cost solutions. For the real-life problem that motivated this study, the proposed approach saves the company an additional 2% over the hierarchical approach, where optimal strategic and tactical decisions are made sequentially. The Benders decomposition solution method with acceleration techniques utilizing disaggregated cuts, dual variables, and the LP relaxation in the initial iterations reduces the running time by a factor of 480, versus a standard Benders decomposition algorithm.

Yilmaz and Catay (2006) consider a variation of PDP involving a single product, multiple suppliers, multiple producers, and multiple distributors, with an option of capacity expansion at additional fixed and variable costs. New continuous variables representing increased capacity, and binary variables indicating capacity expansion decisions for transportation and facility, are introduced. Only manufacturers are allowed to carry inventory, and thus the inventory balance is considered only at the manufacturers' sites. Three different LP relaxation-based heuristics are used to solve the problem, and the relaxed variables are then adjusted to 0 or 1 according to different search mechanisms. The results are then compared with CPLEX solutions obtained with a 300-second time limit.

Another representative study on variables relaxation-based approaches was performed by Lei et al. (2009). The authors considered a single-product multiperiod PDP defined on a $M||B||J$ network with both forward and reverse flows. Because of the need to model the reverse flow in the supply chain network, new constraints such as

$$H_{i,t-1} + \sum_{\forall j} R_{i,j,t} - \sum_{\forall m} R_{m,i,t} = H_{i,t}, \quad \forall i,t$$

are added, where the variable R refers to the reverse flow quantity, and H refers to the reverse product inventory levels. A partial LP relaxation-based rolling horizon method is proposed. With this approach, a given multiperiod planning horizon is partitioned into three intervals: the current period, the immediate next period, and a consolidated period covering all future time periods. In the first interval, all of the original constraints and the integer requirements remain unchanged. For the second and the third intervals, only the integer requirements on the number of truck trips between the DC and customers are relaxed. To reduce the computational effort of each iteration, the forward and backward demands in the third interval are equal to the sum of the forward and backward demands of all the time periods in that interval, respectively. The ending inventories obtained from the solution to the first interval are then fixed as the beginning inventories for the second interval, and this process repeats by redefining intervals until all the time periods achieve integer solutions. Randomly generated test cases are used to benchmark the computational performance of the proposed algorithm against that obtained by the CPLEX within a 1-hour CPU time. More than 70 test cases are randomly generated, and the largest error gap observed is 0.16%, and the required computation time is less than 5 seconds; the average computation time required by CPLEX for solving these cases far exceeds 700 CPU seconds.

3.3.4 Remarks on PDP

In general, if the particular PDP problem being studied has a relatively simple structure, the well-known solution methodologies from the literature can often be effectively adapted. For example, when a PDP problem is defined on a two-stage supply chain network and the constraints are limited to those defined by Equations 3.2 through 3.8, the original problem can be decomposed by either a sequential decomposition or Lagrangean decomposition, which allows the decomposed problem to be modeled as an easy-to-solve problem such as the lot-sizing problem, or a linear programming or network flow problem.

Although not included in this survey, it should be pointed out that in the literature, there has also been a significant amount of work focusing on production and distribution involving uncertainty in demand, processes, and/or supplies, for which stochastic and fuzzy models have been applied extensively. The difference between stochastic and fuzzy models is that a stochastic model usually follows a known probabilistic distribution, while a fuzzy model is described by a simple distribution, such as a triangular distribution, based on expert knowledge. Representative work in stochastic PDP can be found in studies by Park (2005), Aliev et al. (2007), Lejeune and Ruszczynski (2007), and Liang and Cheng (2009). Also note that although the exact methods have rarely been discussed in the literature for solving PDP problems, they could be appropriate if the problem has a special structure, such as that given by Wang et al. (2010).

3.4 The Production and Distribution Problem with Time Constraints

PDP with time constraints (PDPT) is a natural extension of the PDP model, which explicitly takes into account production and transportation time and usually assumes a deadline for the shipment arrival to the customer. To define the shipment arrival times, additional notation must be introduced. Let r_m^k be the production rate for product k at manufacturer m. Let $\tau_{m,i}$ and $\tau_{i,j}$ be the transportation times from manufacturer m to DC i, and from DC i to customer j, respectively. Let $L_{j,t}$ be the deadline at customer site j in period t, by which time the shipment of commodities should have arrived at j; otherwise a shortage or tardiness cost would be incurred. Let MM be a very large positive number. The deadline constraints are defined as follows.

$$\frac{P_{m,t}^k}{r_m^k} + \tau_{m,i} + \tau_{i,j} - \left(3 - Z_{m,t}^k - W_{m,i,t} - W_{i,j,t}\right)MM \le L_{j,t} \quad \forall m,i,j,k,t \quad (3.26)$$

The basic PDPT model is defined by Equations 3.1 through 3.8 and 3.26.

Some papers study PDPT problems involving production lead times and delivery lead times over a multiperiod planning horizon. Let $l_{m,i}$ and $l_{i,j}$ represent lead times from manufacturer m to DC i, and from DC i to customer j, respectively. In this case, Equations 3.2 through 3.4 should be replaced by the following constraints.

$$I_{m,t-1}^k + P_{m,t}^k - \sum_{\forall i} Q_{m,i,t}^k = I_{m,t}^k, \quad \forall m,k,t \quad (3.27)$$

$$I_{i,t-1}^k + \sum_{\forall i} Q_{m,i,t-l_{m,i}}^k - \sum_{\forall i} Q_{i,j,t}^k = I_{i,t}^k, \quad \forall i,k,t \quad (3.28)$$

$$I_{j,t-1}^k + \sum_{\forall i} Q_{i,j,t-l_{i,j}}^k - \left(d_{j,t}^k - S_{j,t}^k\right) = I_{j,t}^k, \quad \forall j,k,t \quad (3.29)$$

Owing to the complexity of PDPT, using a single methodology, such as a Lagrangean relaxation, or a simple heuristic algorithm, may not be effective enough to solve the problem. In the literature, two major approaches have been discussed. One is iteration-based, and starts with an initial solution (or a group of solutions), and then continuously improves the solution (or the set of solutions) iteratively by a relatively simple procedure; most metaheuristic-based algorithms belong to this category. The other is to formulate the original problem into a mathematical model and then use optimization software to derive the optimal or near-optimal solutions. The latter approach has typically been used for solving some case-specific problems.

There are also several papers using simulations to deal with PDPT involving uncertainty. Most such studies (e.g., Lee et al., 2002; Lee and Kim, 2002; Safaei et al., 2010) start with a deterministic version of the problem and solve it to find an initial solution. Through simulation, the solution is evaluated and the parameters of the respective deterministic problem are modified until the solution stabilizes. In this survey, we include only such simulation studies that report on the approaches to solve respective deterministic versions of the PDPT problem.

In this section, we focus on the existing solution methodologies for solving PDPT. Two categories of solution approaches are reviewed: (1) a metaheuristic and iterative approach and (2) mathematical modeling and the use of an optimization solver. Again, we do not consider detailed routing decisions in this section, and hence we treat all transportation operations as direct shipping or fixed routing.

3.4.1 Metaheuristic and Iterative Approach

Naso et al. (2007) consider the integrated problem of finding an optimal schedule for the just-in-time (JIT) production and delivery of ready-mixed concrete with manufacturers and customers. The study involves a single product in a single period with no inventory permitted. Times required for the loading, unloading, and shipping operations of each truck must be explicitly modeled. In addition, outsourcing options of production and third-party (or over-time) trucks are permitted at an additional cost. All decision variables are binary, where $x_{jvr} = 1$ if job j is assigned to truck v as the rth task: $y_{mj} = 1$ if job j is produced at manufacturer m, and $y_{oj} = 1$ if job j is outsourced. The scheduling algorithm combines a GA and a set of constructive heuristics, which are guaranteed to terminate in a feasible schedule for any given set of jobs.

Gebennini et al. (2009) consider a multiperiod strategic and operational planning problem for a single manufacturer that offers a single product with uncertain demand on an $M||B||J$ network. Production lead times and delivery lead times are considered, where lead time may be an integer multiple of one time period, and inventory and stockout costs are considered with safety-stock (SS) determination. Thus, the problem of minimizing the total cost is modeled as a mixed-integer nonlinear programming problem in which the objective function includes a nonlinear term representing the SS cost, $\sum_{i \in B} c_i^s \hat{k} \sqrt{\sum_{j \in J} \hat{\sigma}_{ij}^2 \vartheta_{ij}}$, where c_i^s is the inventory cost for DC i, $\hat{k}$ is a safety factor to control the customer service level, $\hat{\sigma}_{ij}^2$ is the combined variance at DC i serving customer j, and ϑ_{ij} is a 0–1 decision variable equal to 1 if DC i supplies customer j in any time period. This nonlinear term is linearized to

$$\sum_{i \in B} \sum_{j \in J} c_i^s \frac{1}{SS_i} \hat{k}^2 \hat{\sigma}_{ij}^2 \vartheta_{ij}$$ where SS_i is a lower bound on the optimal amount of SS

carried at DC i, because the closer SS_i is to the optimal SS level at DC i, the closer the formula is to the optimal SS cost. A recursive procedure based on the modified linear model is developed to find an admissible solution to the nonlinear model and quantify the minimized global logistic cost, while also taking the effect of safety-stock management into consideration. Because the optimal safety-stock level is unknown, the value is initially set to a lower bound on the effective safety-stock quantity for each DC. It is claimed that the proposed recursive procedure converges to the global optimal solution of the original nonlinear problem in a finite number of iterations.

Yimer and Demirli (2010) address a multiperiod, multiproduct scheduling problem in a multistage build-to-order supply chain manufacturing system with consideration of lead times for production and delivery. For the sake of efficient modeling performance, the entire problem is first decomposed into two subproblems: (1) a downstream part: from manufacturers through distributors and retailers to customers, and (2) an upstream part: from suppliers through fabricators to manufacturers. Both subproblems are then formulated as MIP models with the objective of minimizing the associated aggregate costs while improving customer satisfaction. A GA-based heuristic is proposed with a chromosome of three parts: (1) product ID, total production quantity at each plant, and inventory level at each DC in the period; (2) flow proportion floating values; and (3) status values for feasibility. If a candidate solution is infeasible, it is revised by a proposed repairing heuristic. The fitness value is measured by the original objective function value and the degree of infeasibility. Using some test instances, the best solutions obtained from GA are of high quality compared with the lower bounds obtained from LINGO, a nonlinear programming solver.

Sabri and Beamon (2000) develop an integrated multiobjective supply chain model that facilitates simultaneous strategic and operational planning using an iterative method in a four-tier network. They consider stochastic demand and capacity constraints in all layers of the supply chain, and shortages are allowed, but penalized, while a fixed setup production cost is incurred. Total production lead time at manufacturer m for product k is $g_m^k + \dfrac{Q_m^k}{r_m^k} + l_m^k + \theta_m^k$ where g_m^k, Q_m^k, r_m^k, l_m^k, and θ_m^k are production setup time, production quantity, production rate, waiting time, and material delay time, respectively. θ_m^k is determined by the bill of material of product k and customer service level. They first find a solution for the strategic model and then use the solution as an input to solve the operational model. New parameters determined in solving the operational model are used to solve the strategic model, and this iteration terminates when all binary variables no longer change. LINGO is used in solving each subproblem.

3.4.2 Mathematical Modeling and the Use of the Optimization Solver

Whereas some researchers try to develop effective solution methodologies to solve the PDPT, others put more effort into the modeling process. In this subsection, we summarize the research in which the models are solved by mathematical optimization software such as CPLEX. The common feature of the following papers is that the authors concentrate on the models rather than the design of methodologies. The size of the computational testing instances is small enough for the solver to handle, or else the problem comes from real-world practice so that the solution by a solver is applicable.

Rizk et al. (2006) examined a multiple-product production–distribution planning problem on a single manufacturer and a single destination. The manufacturer operates a serial production process with a bottleneck stage, subject to a predetermined production sequence. The manufacturing cost consists of the changeover cost of intermediate products and the inventory holding cost of final products. The transportation cost is characterized by a general piecewise linear function of transportation quantity with break points of Λ_h with $\Lambda_0 = 0$. In the hth interval $(\Lambda_{h-1}, \Lambda_h]$, let v_h be the slope of its straight line, A_h be the discontinuity gap at the beginning of the interval, and E_h be the ending value. Thus, the transportation cost is $z(\Lambda) = (E_{h-1} + A_h) + v_h \lambda_h$, $\lambda_h = \Lambda - \Lambda_{h-1}$. Valid inequalities to strengthen these formulations are proposed and the strategy of adding extra 0–1 variables to improve the branching process is examined.

Chen and Lee (2004) investigated a multiperiod simultaneous optimization of multiple conflict objectives with market demand uncertainties and uncertain product prices in a supply chain network consisting of manufacturers, DCs, retailers, and customers. The scenario-based approach is adopted for modeling the uncertain market demands, and the product prices are taken as fuzzy variables where the incompatible preference on prices for different participants are handled simultaneously. The whole model becomes a mixed-integer nonlinear programming problem to compromise fair profit distribution, safe inventory levels, maximum customer service levels, and decision robustness to uncertain product demands. Incompatible preference of product prices for all participants is considered by applying the fuzzy multiobjective optimization method; nonlinear MIP solvers, DICOPT and CONOPT, are used for a numerical example.

Dhaenens-Flipo and Finke (2001) provided a multiple period model on an $M||B||J$ network that comes from a practical case at the European industrial division of the manufacturer. Because switching from one product to another on a production line may take a long time, it is assumed that at most one switching per period and per production line is allowed. There are three aggregated products and three line types according to capability to handle these products. All possible sequences in each manufacturing line are enumerated, and they are used in a mixed integer programming model. The set of available product sequences of the line m is denoted by $S(m)$ and these sequences are indexed by s. At this stage, the data involved concern the total production time (B_m) available on line m, the production time $\left(TP_m^k\right)$ and cost $\left(CP_m^k\right)$ of product k on line m, the changeover time (TC_{sm}), and the cost (CC_{sm}) associated with the products of sequence s on line m. Let p_m^k be a quantity of product k manufactured on line m, and let y_{sm} be 1 if sequence s is chosen for the line m. Thus, we need to add the following constraints:

$$\sum_{s \in S(m)} y_{sm} = 1 \quad \forall m \tag{3.30}$$

$$p_m^k - \sum_{s \in S(m):k \in s} y_{sm} \times B_m / TP_m^k \leq 0 \quad \forall m, \forall k \tag{3.31}$$

$$\sum_k p_m^k \times TP_m^k + \sum_{s \in S(m)} y_{sm} \times TC_{sm} \leq B_m \quad \forall m \tag{3.32}$$

The proposed MIP has the constraints in Equations 3.30 through 3.32, flow balance equations similar to Equations 3.2 through 4.4, and domain constraints. For problems of industrial sizes, the model is able to provide a sub-optimal solution in less than 2 hours (23 minutes on the average) by CPLEX.

Fahimnia et al. (2008a) surveyed 20 papers and define a representative mixed integer program formulation for the integration of an aggregate production and distribution plan on an $M||B||J$ network. Three production alternatives are considered: regular-time production, overtime production, or outsourcing. They illustrate with an example to show that considering production alternatives can give a more accurate and better schedule than considering average production cost.

3.4.3 Remarks on PDPT

Lagrangean relaxations and decomposition-based techniques are not effective for solving the general PDPT problems because newly added time constraints often change the model structure significantly. The production and transportation time as well as the incurred deadline constraints all add more complexities to the original PDP, as a feasible solution for a PDP may violate the deadline constraint in PDPT. Even after a PDPT is decomposed, the resulting subproblems may still be NP-hard and therefore make Lagrangean relaxation and decomposition-based solution approaches fail to function effectively. Therefore, most literature results reported are either customized solution approaches for specific PDPTs or efficient algorithms for solving some special cases of PDPT.

3.5 The Production and Distribution Problem with Routing

PDP with routing (PDPR) is discussed in this section. Because of its complex structure, most papers assume a two-stage network, and thus those problems can be considered as a combination of the capacitated lot-sizing problem and the inventory routing problem. The aim of the problem is to minimize the total cost, composed of inventory holding, production, and transportation costs.

We consider a basic model defined on a two-echelon supply chain consisting of a set of manufacturers and a set of customers, where customer j has demand d_{jt} in period t. For simplicity, a single product is considered and thus the superscript for product type (k) is dropped. We assume that there is a fleet of homogeneous vehicles belonging to manufacturer m, denoted by $V(m)$. Because the PDPR model contains routing decisions in it, the quantity being carried by a vehicle is different from the quantity delivered to a customer by a vehicle in a period. Thus, the following parameters and decision variables are added to PDP:

f_{mjlt}^v = fixed cost of vehicle v of manufacturer m along (j, l) in period t

g_{mjlt}^v = unit shipping cost for vehicle v of manufacturer m along (j, l) in period t

ξ_{mjlt}^v = equals 1 if vehicle v of manufacturer m serves l immediately after j in period t

Q_{mjlt}^v = quantity carried by vehicle v of manufacturer m along (j, l) in period t

q_{mjt}^v = quantity delivered by vehicle v of manufacturer m to customer j in period t

for $m \in M, j \in \{m\} \cup J, l \in \{m\} \cup J, t \in T$.

The objective function of the model consists of production, inventory, and transportation (routing) costs. The transportation cost is changed as follows:

$$\sum_t \sum_m \sum_{v \in V(m)} \sum_{j, l \in \{m\} \cup J, j \neq l} f_{mjlt}^v \xi_{mjlt}^v + g_{mjlt}^v Q_{mjlt}^v \qquad (3.33)$$

Moreover, routing constraints should be included in the model. The flow conservation constraints are

$$\sum_{l \in \{m\} \cup J, l \neq j} Q_{mjlt}^v - \sum_{l \in \{m\} \cup J, l \neq j} Q_{mljt}^v = -q_{mjt}^v \quad \forall m, v, j, t \qquad (3.34)$$

$$\sum_{j \in J} Q_{mjmt}^v - \sum_{j \in J} Q_{mmjt}^v = -\sum_{j \in J} q_{mjt}^v \quad \forall m, v, t \qquad (3.35)$$

We need an inventory balance constraint for each customer.

$$I_{j,t-1} + \sum_{m} \sum_{v \in V(m)} \sum_{j \in J} q_{mjt}^v - d_{jt} = I_{j,t} \quad \forall j,t \tag{3.36}$$

Because ξ_{mjlt}^v represents the existence of flow on (j, l) and each customer can be served by at most one manufacturer, we have the following constraints:

$$Q_{mjlt}^v \leq \xi_{mjlt}^v MM \quad \forall m,v,t,j,l \in J \tag{3.37}$$

$$\sum_{m} \sum_{v \in V(m)} \sum_{l \in \{m\} \cup J, l \neq j} \xi_{mljt}^v \leq 1 \quad \forall j,t \tag{3.38}$$

We classify the relevant papers, according to their solution methodologies, into three classes. Because the problem includes the routing decisions, all methods use decomposition. However, each decomposed problem is solved by a different solution approach. One approach is to use mathematical programming or simple heuristic algorithms. The other two use a metaheuristic, such as a tabu search, and the approximation approach, respectively.

3.5.1 Mathematical Programming Approach

Fumero and Vercellis (1999) studied a multiple period and multiple product problem with a single manufacturer. They assumed that there are fixed setup costs and vehicle usage costs that occur independently from the amount of produced or carried product. In the model, partial order serving is allowed. They decompose the problem into production (capacitated lot-sizing) and distribution (multiperiod vehicle routing) problems by Lagrangean relaxation, relaxing the constraints that ensure the balance at the central plant among production, inventory, and deliveries. Furthermore, the vehicle capacity constraints are relaxed to simplify the solution of the routing subproblem. The Lagrangean dual problem is solved using a variable target subgradient optimization algorithm that is described in Fumero and Vercellis (1997). In addition, they employ an alternative decomposition method in which the production plan is developed without considering the distribution plan, and then used as an input for the distribution model. They show that the Lagrangean decomposition method outperforms the alternative decomposition method.

Bard and Nananukul (2010) proposed a hybrid methodology that is a combination of an exact method and heuristic procedures within a branch-and-price (B&P) framework for the problem with a single manufacturer and a single product type. The master problem (MP) is defined by the production and inventory decisions, and the remaining routing problem can be decomposed by period, yielding $|T|$ subproblems. In the reformulated model, each column in the MP corresponds to a feasible schedule for all customers. They use a novel column generation heuristic and a rounding heuristic to improve the algorithmic efficiency. They show that the B&P heuristic is efficient and can derive high-quality solutions for large problems within a reasonable amount of time.

Ruokokoski et al. (2010) considered the problem of determining a production schedule for an uncapacitated plant, replenishment schedules for multiple customers, and a set of routes for a single uncapacitated vehicle. The aim of the problem is to fulfill customer demand over a finite horizon at a minimum total cost of distribution, setups, and inventories. This paper introduces a basic mixed integer linear programming formulation and provides exact methods through several strong reformulations of the problem. Moreover, two families of valid inequalities, 2-matching and generalized comb inequalities, are introduced to strengthen these formulations, and they are used within a branch-and-cut framework. Comb inequalities are known to be facets defined for the traveling salesman problem (Grötschel and Padberg, 1979) and 2-matching inequalities are generalized comb inequalities under certain conditions. An *a priori* tour-based heuristic is also provided, and with available solvers and strong formulations, excellent solutions can be obtained within a short time, even for the largest problems.

Archetti et al. (2011) considered a production-routing system, where a manufacturer with unlimited capacity produces one product, which is distributed to a set of retailers by a fleet of vehicles. The objective is to determine the production policy, the customer replenishment policy, and the transportation policy so that the system cost is minimized. Two types of replenishment policies are studied: maximum level (ML) and order-up to level (OU), and the problem is NP-hard under both. As the problem with OU policy has been solved heuristically by Bertazzi et al. (2005), the authors proposed a three-step sequential heuristic on the ML policy. In the first step, unlimited production quantity is assumed, and the distribution part of the problem concerning inventory at customers and delivery routes is optimized by solving a customer problem with branch-and-cut and iteratively adding it to the solution. In the second step, the production plan is determined by solving the classical uncapacitated lot-sizing problem, which can be optimally solved in polynomial time. In the third step, the improvement procedure, removing and inserting two retailers, is repeated until there is no further improvement.

Cetinkaya et al. (2009) considered a three-layer practical supply chain problem and developed a multiproduct and multiperiod model to improve the outbound supply chain of Frito Lay North America (FLNA), consisting of a factory warehouse, multiple DCs, and a set of customers. Some customers can receive supplies directly from the factory warehouse, which is called direct delivery (DD). They did not consider the production costs but the production capacities. The objective function includes the inventory holding cost, the truck loading and dispatch cost, mileage costs, and handling costs. The proposed solution methodology decomposes the integrated problem into two subproblems—inventory and routing problems—and they are iteratively solved until either no further improvement is found or the maximum number of iterations is reached. The routing subproblem is solved period by period. As a preprocessing, they use full-truck load (FTL) shipments with a route having a single destination for customers with large order quantities, and use less-than-truck load (LTL) shipments with truck routes for other customers. They then use a savings algorithm proposed by Clarke and Wright (1964) and utilized by Chopra and Meindl (2001) and add an improvement step, called the cheapest insertion heuristic, a well-known traveling salesman problem heuristic. For the inventory subproblem, the objective function includes the corresponding route-based setup costs and all cost terms of the overall model, except the loading and routing parameters considered in the routing subproblem. The CPLEX 9.0 solver is used to solve the inventory subproblem.

3.5.2 Metaheuristic Approach

Bard and Nananukul (2009a) consider the problem of a $B||J$ network in which inventory handling at both the customer and manufacturer sites is permitted, but the inventory level must be zero at the end of the each period, with no shortages allowed. They solved the problem using a two-phase approach, which is similar to the method developed by Lei et al. (2006). In the first phase, they formulated the model as a mixed integer program without taking into account the routing constraints. They found a feasible solution that determines the sufficient delivery amounts for all customers using the proposed model. The solutions derived in the first phase are used as an initial solution for the tabu search algorithm, which is used in the second phase to solve the integrated problem. The path relinking method is used to obtain better solutions. They showed that the lower bounds obtained from the relaxed version in the first phase are not very effective for evaluating the proposed algorithm. However, according to the computational results, the proposed method can derive 10%–20% better solutions, but requires more computational effort than the greedy randomized adaptive search procedure (GRASP) proposed by Boudia et al. (2007).

Bard and Nananukul (2009b) proposed three algorithms with a B&P framework for the Inventory Routing Problem (IRP) as a subproblem of the integrated production–inventory–distribution–routing problem. For less computation, a two-step procedure is proposed: the first step involves developing a model for determining delivery quantities for each customer in each period. The second step involves finding actual routes in light of the current set of branching constraints with a vehicle routing problem (VRP) tabu search method. According to computational experiments, although the B&P algorithm generates better results than the tabu search approaches (3.6% on average), the tabu search outperforms the B&P algorithm in terms of the computation time (more than 10 times faster on average).

Yossiri et al. (2012) developed a decomposition heuristic based on an adaptive large neighborhood search (ALNS) for the problem defined on a network consisting of a plant and multiple customers to minimize the total production, setup, inventory, and routing costs. In the first stage, a set of initial solutions are generated with different setup schedules by solving two subproblems: (1) production and distribution problem with approximate transportation costs and (2) routing problem; both are solved heuristically. In the second stage, the initial solutions are improved by ALNS. When a solution is modified by removing a customer from a route and inserting it in a different period, one has to identify the new delivery quantity for the customer, which may also affect the production, inventory, and other delivery quantity decisions. It is not always necessary to reinsert the removed nodes, because the demands can be satisfied from available inventory and, furthermore, the removed nodes can be inserted in multiple periods. To deal with these issues, binary variables are defined accordingly. During the transformation process, the binary decisions concerning routing are modified according to the cheapest insertion rule and then, with fixed binary variables concerning production setup, the continuous variables are adjusted by solving the minimum cost flow problem.

3.5.3 Incorporating Routing Cost Approximation for Solving PDPR

When the decomposition method is applied, a PDPR problem is usually solved through two phases. During the first phase, a reduced version of PDPR is solved, where many studies assume direct shipments to customers (e.g., Lei et al., 2006); and then during the second phase, vehicle routing decisions are made to improve the solutions obtained in the first phase. The advantage of such a phased approach is to reduce the search complexity in each phase. However, using direct shipment to replace vehicle routing in the first phase can sometimes also lead to a solution that is feasible but deviates significantly from the optimal solution to the original problem.

Another alternative solution approach to PDPR is based on continuous approximation models for the vehicle routing problems. Such an approach uses a continuous approximation of the optimal routing cost in the phase I problem instead of assuming direct shipments. Note that this provides an estimation of the actual routing cost without explicitly solving the vehicle routing problem. Once the phase I problem is solved and the assignments of vehicles to customers are determined, the exact routing decisions under the given vehicle assignments are made during the second phase.

Shen and Qi (2007) incorporated a continuous approximation function in their integrated supply chain design model to estimate the optimal vehicle routing cost. Specifically, the approximate function that they propose is

$$V_{mvt} = \frac{2}{C^v} \sum_{j \in J} q_{mjt}^v \mu_{mjt}^v + \left(1 - \frac{2}{C^v}\right) \Phi |v_t| \sqrt{\frac{A}{|J|}}$$

where

V_{mvt} = the approximate routing cost of vehicle v of manufacturer m in period t

C^v = the capacity of vehicle v of manufacturer m

q_{mjt}^v = the quantity delivered by vehicle v of manufacturer m to customer j in period t

μ_{mjt}^v = the unit cost of a direct shipment by vehicle v of manufacturer m to customer j in period t

$|v_t|$ = the number of customers served by vehicle v of manufacturer m in period t

A = the area where customers are scattered

Φ = parameter, and $\Phi = 0.75$ for Euclidean metrics

Shen and Qi (2007) numerically demonstrated the effectiveness of this approximation function using a data set with 150 points from Christofides et al. (1979), and showed that this approximation function performs especially well when the number of customers is sufficiently large. In particular, when the number of customers is more than 80, the approximation error is typically less than 5%.

When the above continuous approximation function is incorporated in the phased approach, parameters q_{mjt}^v and $|v_t|$ vary with the decision of assignments of vehicles to customers, while all the remaining parameters are given constants. Compared with the direct shipment assumption that is often made in the literature, this approximation function provides a more accurate estimation of the routing cost without increasing the problem complexity. This approach may be used as an alternative to further enhance the performance of phased approaches.

3.5.4 Remarks on PDPR

In this section, the total cost of the PDP with routing is minimized, where the total cost is composed of inventory holding, production, and routing costs. Because the problem includes the vehicle routing problem, it is very difficult to find the optimal solution or an approximate solution close to the optimum.

Thus, most algorithms use a decomposition approach and metaheuristic algorithms, such as a tabu search, to solve routing subproblems. When there is a single manufacturer, the decomposition approach is frequently used because the upstream problem can be regarded as a capacitated lot-sizing problem. Moreover, after obtaining a solution, various improvement heuristics are also often used as post-processing procedures. Because the optimal value is usually unavailable, the performance of an algorithm is presented by comparing its solution with a lower bound, or with a solution obtained by either previous approaches or an optimization solver.

3.6 The Production and Distribution Problem with Routing and Time Constraints

PDP with routing and time constraints (PDPRT) is discussed in this section. Time constraints appear in different forms, such as time window, due date, and exact arrival time predetermined by customers. It can be considered as a combination of an inventory routing problem (IRP) with time constraints and a capacitated lot-sizing problem.

In most of the existing literature, a two-echelon supply chain that contains a single plant and a set of geographically scattered customers is considered. Because of the complexity of the problem, multiple manufacturers are rarely considered (see Lei et al., 2006; Bilgen and Günther, 2010). Generally, the objective function contains the production cost, the transportation cost (routing cost) and the inventory holding cost. On the other hand, minimizing the makespan consisting of production time and transportation time, and maximizing the satisfied demand are considered as objectives in Geismar et al. (2008) and Armstrong et al. (2008), respectively. Although third-party vehicles are rarely considered, Lei et al. (2006) take third-party transshipments into account. The authors consider a representative model with a two-echelon supply chain network consisting of a set of manufacturers and a set of customers. For simplicity, they assume that customer j has demand d_{jt} with due date L_{jt} in period t, and that there is a fleet of homogeneous vehicles belonging to each manufacturer. To deal with time constraints, additional parameters and variables are defined. The objective function and constraints other than time constraints are equivalent to those in the model in Section 3.5. Thus, focusing only on time constraints:

τ^v_{mjlt} = travel time of vehicle v of manufacturer m on arc (j, l) in period t

T^v_{mjt} = arrival time of vehicle v of manufacturer m at customer j in period t

To guarantee due date constraints, the authors add the following constraints:

$$T_{mjt}^v + \tau_{mjlt}^v \le T_{mlt}^v + MM\left(1 - \xi_{mjlt}^v\right) \quad \forall m, v, t, j, l \in J, j \ne l \quad (3.39)$$

$$T_{mlt}^v \le L_{lt} \quad \forall m, v, t, l \in J \quad (3.40)$$

The solution methodologies used to solve this problem in the literature fall into two different groups, according to their structures; the first one solves the problem in an integrated manner, while the second one partitions the problem into small pieces that are easier to solve. In these decomposition methods the solution from the first phase is used as an input to the second phase. Using integrated methods, the solution may be improved by an iterative process.

Chandra and Fisher (1994) solve the production and transportation scheduling problems in separate and integrated manners and compare those results. In their model, the plant can produce several products in a limited time and transporters are allowed to partially deliver to a set of customers with unlimited capacity in a period. The plant has an unlimited production capacity and the inventory holding costs are not involved in the total costs. First, they implement an integrated approach in small examples and show that firms can reduce their operation costs about 3%–20% by coordinating the production and distribution activities. Second, in the decomposed part, they assume that the production scheduling problem can be modeled as a capacitated lot-sizing problem and the distribution problem can be modeled as a standard multiperiod local delivery routing problem. The interface of GAMS, ZOOM/XMP, a solver, is used to solve the production scheduling problem. They use three well-known vehicle routing heuristics—sweep (Gillett and Miller, 1974), nearest neighbor rule (Rosencrantz et al., 1974), and feasible insertion rule (Chandra, 1989)—to find an initial solution to the distribution problem. A local improvement heuristic is used to combine the production and distribution problems. Since the work of Chandra and Fisher (1994), many extended studies have been conducted with various approaches including decomposition and compounded methods.

3.6.1 Decomposition Methods

Using decomposition methods, the problem is usually partitioned into two subproblems—production planning and routing problems—that are solved sequentially.

Lei et al. (2006) investigated an integrated production, inventory and distribution routing problem where there is no fixed cost of using a vehicle, and each transporter can make multiple trips during each period. They used a two-phased approach that solves the problem in two separate stages but in an integrated manner. In the first phase, they assumed that the distribution of the products from plants to customers is carried out by direct shipment. The problem is formulated as a mixed integer programming problem, neglecting the vehicle routing constraints, and solved by the CPLEX MIP solver. In the second phase, they propose a heuristic transporter routing algorithm, called the load consolidation (LC) algorithm, to consolidate the loads into routing decisions. The LC algorithm determines the sequence of transporter trips and allocates the transporters to the trips without violating the transporter capacity and available time constraints. The extended optimal partitioning (EOP) procedure is used to find the shortest path among the feasible trips that are identified in the first phase. They compare the LC algorithm and CPLEX MIP solver with 56 test problems. According to their test results, the LC algorithm can solve the problem in less than 0.2 second whereas the CPLEX MIP solver needs more than 2 hours to solve the overall problem.

Geismar et al. (2008) developed a two-phase heuristic to solve a single-period integrated production and transportation scheduling problem for a product with a short life span. The first phase uses either a genetic algorithm (GA) or a memetic algorithm (MA) to select a locally optimal permutation of a given set of customers. MAs have a local search parameter and a relatively small population size as a result of different population management. In the second phase, for a given permutation of customers, Beasley's (1983) "first route-cluster second" method is used to determine simultaneously the customers to be served and the vehicle routes to be used, and a linear program formulation is used to minimize the makespan for a given set of trips. The Gilmore–Gomory (1964) algorithm for two-machine no-wait flowshops is then used to order the subsequences of customers to form the integrated schedule.

3.6.2 Integrated Methods

Among the papers dealing with integrated methods, some papers propose problem-specific methodologies for problem solving, while others focus on new modeling techniques.

Boudia and Prins (2009) examined a multiperiod production distribution problem in a two-echelon supply chain that is very close to the model proposed by Chandra and Fisher (1994), but differs in that the limited vehicle capacity and a single product are considered. They use a memetic algorithm with population management (MA/PM) to handle production and distribution problems simultaneously. The proposed algorithm is evaluated in three sets of 30 instances with 50, 100, and 200 customers over 20 periods. They compared the proposed algorithm with two previous algorithms: the two-phase algorithm (H1) proposed by Boudia et al. (2005) and the three-phase algorithm (H2) based on GRASP developed by Boudia et al. (2007). They showed that the memetic algorithm can generate better solutions than GRASP, which also solves the related problem from an integrated perspective.

Armstrong et al. (2008) solved a similar problem with a branch-and-bound search procedure to maximize the total satisfied demand by choosing a subset of customers from the given sequence who will be served by a single vehicle. The constraints of the problem refer to the product lifespan, the production/ distribution capacity, and the delivery time window. Because there is no inventory handling at the supply chain members, it is important to synchronize the production and distribution planning decisions successfully. Empirical studies on the computational effort required by the proposed search procedure compared with that required by CPLEX on randomly generated test cases are summarized. A branch-and-bound search algorithm is also proposed and is shown to outperform CPLEX with limited running time.

Bilgen and Günther (2010) considered an integrated production and distribution planning problem in the fast-moving consumer goods industry, with a so-called block-planning approach, which establishes cyclical production patterns defined by setup families. The aim is to minimize the total cost, consisting of production costs, inventory holding costs at distribution centers, and transportation costs for FTL and LTL transportation modes. Unlike the other related studies, they considered two types of production setup costs—major setup costs for each block started on one of the lines (e.g., for clean-out in the food industry) and minor setup costs for the production lots of the individual products. They traced the time in terms of the block and lot production completion times. Two different periods were used in this study: macro periods (e.g., weeks) were used for the block assignments and micro periods (e.g., days) were used for the distribution schedule and external demand elements. They compared two different block-planning approaches: the flexible and the rigid block, which differ by their degree of flexibility in the scheduling of the production lots. A mixed-integer linear programming model is proposed and CPLEX is used as a solver. The numerical results reveal that the flexible block-planning approach can provide considerable cost savings compared with the rigid block-planning approach.

Bolduc et al. (2010) considered the split delivery vehicle routing problem with production and demand calendars. They propose a simple decomposition procedure to provide a starting solution and use a tabu search with new neighbor reduction strategies. After the tabu iterations are completed, an improvement heuristic is applied. They implement their procedure on a randomly generated 100 instances with 50 customers and 10 periods. The results show that the developed model is effective in terms of both solution quality and computation time.

3.6.3 Remarks on PDPRT

In the decomposition method, there are two general approaches: the first one considers the production problem and the routing problem separately, while the second solves the problem including production and simplified distribution and then solves the routing problem. For the integrated method, there are two approaches. The first is to solve the problem simultaneously using mathematical programming with an optimization package, while the second is to use an iterative method in which the solution is improved over iterations through a metaheuristic such as GA and tabu search. Even though there is no clear dominance between the decomposed method and the composed method, the decomposed method is always useful to find an initial feasible solution. For example, Bolduc et al. (2010) used a decomposed method to find an initial solution and then improved it by a tabu search algorithm in an integrated manner.

Because the problem is already complicated by including the vehicle routing problem in it, researchers have focused on a two-echelon problem with static demand. Thus, natural generalizations are required, such as two echelons to multiple echelons, static demand to stochastic demand, and excluding third party to including third party.

3.7 Discussion

In a realistic situation, such as multiproduct, multiechelon, distribution routing, the problem under consideration has a complicated structure with a huge size. Moreover, each problem in the literature has its unique assumptions and definitions. Various approaches are considered and analyzed for different problems, and therefore it is very difficult to propose an integrated view of the entire set of methodologies. In this section, we provide three different perspectives. The first one classifies the solution approaches with a perspective on the decomposition framework, and solution methodologies applied to the decomposed subproblems. The second one relates the problem structure to the utilized solution approaches. The last one addresses future research directions.

3.7.1 Structure of Solution Approach

Most problems in the literature are computationally difficult to solve optimally, and thus different decomposition approaches are utilized. When the problem is decomposed, the optimality of the problem may not be guaranteed, but each decomposed problem is much easier to solve and sometimes can be solved effectively (e.g., optimally or near-optimally) and efficiently (e.g., in polynomial time or in pseudo-polynomial time). Moreover, after the original problem is decomposed into subproblems, each subproblem can be further decomposed according to the structure of the subproblem. The overall framework of the solution methodology in terms of decomposition has the following three categories.

No decomposition. The entire problem is solved at once.

Mathematical decomposition. The original problem is decomposed according to mathematical properties. Two representative decompositions are Lagrangean decomposition and Benders decomposition. In Lagrangean decomposition, some of constraints are relaxed by Lagrangean relaxation and the problem under consideration can be decomposed into independent subproblems. In Benders decomposition, some of the variables are fixed and the problem can be decomposed.

Heuristic decomposition. The original problem is decomposed according to problem-specific properties. A common way is to decompose the problem with respect to layers. Thus, the upstream problem and the downstream problem are separately defined. Another method is to decompose into a strategic problem and an operational problem.

When the problem (or decomposed subproblem) cannot be further decomposed, or is going to be solved directly, several approaches are utilized. The major solution approaches in the literature can be summarized:

Exact algorithm development. When the problem (or subproblem) can be formulated as a problem that has a known optimal algorithm in polynomial time (or pseudo-polynomial time), it can be solved optimally. Typical examples are network flow problems, linear programming (LP), and dynamic programming.

Modeling with an optimization solver. Some papers describe the problem with an exact mathematical formulation, such as LP, nonlinear programming (NLP), and mixed integer programming (MIP), and solve it with an optimization solver. When the problem size is small enough or the problem has unique properties, optimal solutions can be obtained in a reasonable time frame. Various optimization solvers are found in the literature, such as CPLEX, GAMS, AMPL, LINGO, and GLPK. To strengthen the formulation, additional constraints, such as valid inequalities, can be inserted. In most cases, an approximate solution by an optimization solver is acceptable, given the error limit or running time limit.

Mathematical programming approach. When the subproblem is still too hard to be solved optimally, there are several approaches utilizing mathematical programming techniques. Representative methods are Lagrangean relaxation and LP relaxation.

Metaheuristic. Metaheuristics iteratively improve a candidate solution with regard to a given measure of quality. A metaheuristic makes few or no assumptions about the problem being optimized and can search very large spaces of candidate solutions. However, it does not guarantee that an optimal solution is ever found. The solution quality and running times are highly dependent on the setup parameters. Examples are local search (e.g., tabu search, simulated annealing), evolutionary algorithms (e.g., genetic algorithm), and swarm intelligence (e.g., particle swarm optimization, ant colony optimization).

Problem-specific algorithms. According to the problem-specific property, an algorithm can be developed only for the particular problem. In many cases, values of variables are sequentially decided. A representative one is a greedy algorithm, which makes a locally optimal choice at each stage with the hope of finding a global optimum. After obtaining a solution, a local improvement procedure may be applied.

Figure 3.2 gives an overview of the existing procedures for solving the integrated problem. If a problem is directly solvable, it can be solved using an exact method. Otherwise, we may try to decompose it into multiple subproblems with minor changes from the original problem, or try to use other solution approaches. If the problem is decomposed, subproblems can be solved separately and each of them is considered as an independent problem. Then, we can iteratively check whether the subproblems are directly solvable or further decomposable. If the problem (or subproblem) is not decomposable or we do not attempt to decompose it further, several solution approaches are applicable.

Based on the above classification, the solution approaches used in the literature surveyed in this chapter can be classified in Table 3.2. We make the following observations.

When the problem is solved without decomposition, the two major methodologies are modeling with an optimization solver and a metaheuristic, in which the structural property is not well utilized.

When a mathematical decomposition is utilized as an overall framework, the subproblem is always solved by mathematical programming methods for optimal or approximate solutions. In other words, if one would like to apply mathematical decomposition, subproblems should be able to be well handled by mathematical programming methods.

When the problem is heuristically decomposed, metaheuristic and problem-specific heuristics are frequently used.

3.7.2 Problem Structure and Solution Approaches

In the reviewed papers, along with the problem structure and methodologies used, when routing is involved as a part of the decision, the problem includes a vehicle routing problem (VRP), which is one of the well-known difficult combinatorial optimization problems. Thus, we separately discuss the problems where routing is considered, and those where it is not.

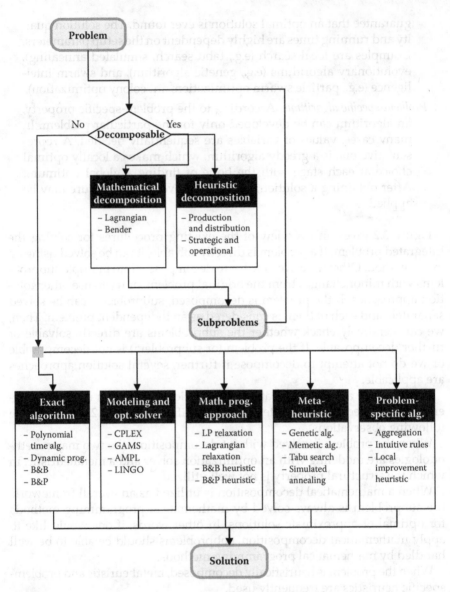

FIGURE 3.2
An overview of existing procedures for solving the integrated problem.

For the problems without routing decisions (PDP and PDPT), the methodologies for PDP and PDPT differ.

The major solution methodology for PDP is to use Lagrangean decomposition as a framework and mathematical programming for the decomposed problems. Especially when the PDP is defined on a supply chain network with two stages, Lagrangean decomposition works very well, because the

TABLE 3.2

Summary of Solution Approaches

Overall Framework	No Decomposition	Mathematical Decomposition	Heuristic Decomposition
Subproblem methodology			
Modeling with optimization solver	Rizk et al. (2006) Chen and Lee (2004) Dhaenens-Flipo and Finke (2001) Fahimnia et al. (2008a) Bilgen and Günther (2010)		Sabri and Beamon (2000) Cetinkaya et al. (2009) Chandra and Fisher (1994)
Exact algorithm development	Armstrong et al. (2008) Ruokokoski et al. (2010)	Yung et al. (2006) Eksioglu et al. (2007) Karakitsiou and Migdalas (2008) Dogan and Goetschalckx (1999)	Bard and Nananukul (2009b) Archetti et al. (2011)
Mathematical programming approach	Yilmaz and Catay (2006) Lei et al. (2009)	Fumero and Vercellis (1999)	Bard and Nananukul (2010) Bard and Nananukul (2009b) Archetti et al. (2011)
Metaheuristic	Jang et al. (2002) Gen and Syarif (2005) Kannan et al. (2010) Naso et al. (2007) Boudia and Prins (2009) Bolduc et al. (2010)		Ahuja et al. (2007) Yimer and Demirli (2010) Bard and Nananukul (2009a) Geismar et al. (2008) Yossiri et al. (2012)
Problem-specific algorithm	Lei et al. (2006) Liu et al. (2008) Gebennini et al. (2009) Shen and Qi (2007)		Park (2005) Cetinkaya et al. (2009) Chandra and Fisher (1994) Lei et al. (2006) Geismar et al. (2008) Archetti et al. (2011)

subproblems can be solved optimally. However, when PDP is defined on a network with three or more stages, Lagrangean decomposition is rarely used.

The major methodology of PDPT is to establish a mathematical model without decomposition and use an optimization solver. Half of the papers dealing with PDPT use an optimization solver, even though some mathematical models are nonlinear, while no papers use mathematical programming for overall or decomposed problems. It may imply that the problem with time constraints can be clearly defined in a mathematical model, but the time constraints make it difficult to utilize the mathematical structure for mathematical programming-type algorithm development.

For the problems with routing decisions (PDPR and PDPRT), mathematical decomposition is rarely used, while heuristic decomposition is frequently used. When the problem is decomposed heuristically, the upstream problem deals with production lot-sizing and the downstream problem is defined for routing decisions. Decomposed subproblems are solved by various methods.

In PDPR, one subproblem may be modeled and solved by an optimization solver, and the other subproblem solved by a problem-specific heuristic. In another case, one subproblem is solved by mathematical programming for an approximate solution, while the other subproblem is solved by an exact algorithm for the optimal solution.

In PDPRT, a mathematical programming approach is rarely used as the solution methodology for decomposed problems because of the complexity of the decomposed problems. Instead, metaheuristic and problem-specific heuristic approaches are widely used. In both PDPR and PDPRT, the solution approaches cannot directly give a solution close to the optimum, and thus local improvement heuristics are frequently used as a post-processing procedure.

In addition, we observe the following relationships between problem structure and methodologies used.

The mathematical programming approach works better for problems without time constraints.

When the problem structure is complicated, problem-specific algorithms and local improvement heuristics are frequently used.

Metaheuristics can be applied for most problem structures.

3.7.3 Future Research Directions

One trend in solution approaches for modern supply chain operations, which is much more complex than those addressed in traditional operations related literature, is to use a hybrid methodology by combining the aforementioned methods and the use of a simulation as a framework, especially for practical and large-scale problems. When a simulation is used as a framework for solving the problem, a mathematical model is usually established first and solved as a deterministic problem (by fixing the values of uncertain factors). The resulting solution is then used as an input to the simulation model, which incorporates the uncertainties in demand, facility failure, delivery time, capacity, and so forth. The output of a simulation model helps to adjust the values of model parameters. Such a simulation procedure is usually repeated many times until the solution is efficient and robust. In the current literature, however, there remains a significant dearth of studies that focus on the development and applications of such a comprehensive and hybrid solution methodology.

Another observed trend is the construction of a general framework to deal with an integrated problem of a practical size. Each component of the framework should be separately modeled and possible solution methodologies should be proposed; the decision process, including information granularity and the decision period, should be carefully designed. We may need to consider more qualitative decisions beyond the total cost. A big company may prefer amicable small companies as its partners even though they are not currently cost effective. Alternatively, an industry-dependent framework is a possible direction. For example, the framework for the supply chain network in electronics manufacturing might give rise to a general framework to handle various operations-level decisions.

Furthermore, in most of the literature, we find that stochastic factors are seldom incorporated in the models with time constraints and routing issues, although most real-life integrated production and distribution problems often contain stochastic factors. Also, most papers dealing with routing issues only consider a supply chain with at most two or three echelons. A third party is generally not considered, and when it is considered, it usually has unlimited or very large capacity and zero or very short lead time. Thus, future research may be directed toward an extension of the models to cover more general cases.

Last, but not least, most of the work in the existing literature applies to a collaborative environment, where all information is shared and a decision is made by a central authority and applied to all players in the supply chain. Realistically, this is not always the case. In practice, in most supply chains, each player (or group of players) may pursue its own objective and not all players try to achieve a global optimum. Even where some or all may try to collaborate, information-sharing may not be possible. Each entity may be subject to a different management policy in terms of a given information item, sharing scheme, updating period, and so forth. However, in the current literature, there are few studies that focus on multiechelon supply chain networks with partial information. Although a competitive supply chain has been studied in the past decade, researchers have generally assumed only two echelons, either suppliers and manufacturers, or manufacturers and retailers. In the case where competition exists across the entire supply chain, or in a part of the supply chain, the global optimum may differ significantly from a local optimum. Thus, a game-theoretic approach to modeling and solving competitive but integrated supply chain problems may also be a promising area for future research.

In summary, the gaps between modern supply chain operations problems and problems in the literature can be characterized as (1) large scale (e.g., multiechelon, third party); (2) multiobjective (e.g., total cost, ease of operations); (3) uncertainty (e.g., randomness, disruption); and (4) competitive environment with incomplete information (e.g., multiagents, competing entities). One possible way to incorporate these characteristics is by using simulation as a framework. A large-scale problem can be decomposed into multiple modules such that each module includes collaborative entities. For decision making within a module, the integrated production/distribution/routing

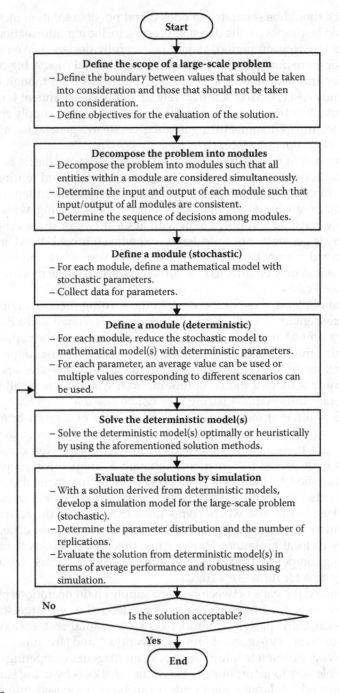

FIGURE 3.3

A general procedure to define and solve modern supply chain operations problems with a simulation framework.

planning problem is defined and solved by various solution methodologies; for decision making among competing modules, game-theoretic analysis can be applied. Multiple objectives can be simultaneously evaluated from different solutions for supply chain networks with uncertainties. Figure 3.3 shows a general procedure for defining and solving these problems within a simulation framework.

References

Ahuja, R. K., T. L. Magnanti, and J. B. Orlin. *Network Flows: Theory, Algorithms, and Applications*. Englewood Cliffs, NJ: Prentice Hall, 1993.

Ahuja, R. K., W. Huang, H. E. Romeijn, and D. R. Morales. A heuristic approach to the multi-period single-sourcing problem with production and inventory capacities and perishability constraints. *INFORMS Journal on Computing*, 19(1): 14–26, 2007.

Aliev, R. A., B. Fazlollahi, B. G. Guirimov, and R. R. Aliev. Fuzzy-genetic approach to aggregate production–distribution planning in supply chain management. *Information Sciences*, 177(20): 4241–4255, 2007.

Archetti, C., L. Bertazzi, G. Paletta, and M. G. Speranza. Analysis of the maximum level policy in a production-distribution system. *Computers and Operations Research*, 38(12): 1731–1746, 2011.

Armstrong, R., S. Gao, and L. Lei. A zero-inventory production and distribution problem with a fixed customer sequence. *Annals of Operations Research*, 159(1): 395–414, 2008.

Bard, J. F. and N. Nananukul. The integrated production–inventory–distribution–routing problem. *Journal of Scheduling*, 12(3): 257–280, 2009a.

Bard, J. F. and N. Nananukul. Heuristics for a multi period inventory routing problem with production decisions. *Computers and Industrial Engineering*, 57(3): 713–723, 2009b.

Bard, J. F. and N. Nananukul. A branch-and-price algorithm for an integrated production and inventory routing problem. *Computers and Industrial Engineering*, 37(12): 2202–2217, 2010.

Beasley, J. E. Route-first cluster-second methods for vehicle routing. *Omega*, 11(4): 403–408, 1983.

Bertazzi, L., G. Paletta, and M. G. Speranza. Minimizing the total cost in an integrated vendor-managed inventory system. *Journal of Heuristics*, 11: 393–419, 2005.

Bilgen, B. and H.-O. Günther. Integrated production and distribution planning in the fast moving consumer goods industry: A block planning application. *OR Spectrum*, 32(4): 927–955, 2010.

Bolduc, M.-C., G. Laporte, J. Renaud, and J. J. Boctor. A Tabu search heuristic for the split delivery vehicle routing problem with production and demand calendars. *European Journal of Operational Research*, 202(1): 122–130, 2010.

Boudia, M. and C. Prins. A memetic algorithm with dynamic population management for an integrated production–distribution problem. *European Journal of Operational Research*, 195(3): 703–715, 2009.

Boudia, M., M. A. O. Louly, and C. Prins. Combined optimization of production and distribution. In *International Conference on Industrial Engineering and Systems Management (IESM'05, Marrakech, Morocco)* Mons, Belgium, 2005.

Boudia, M., M. A. O. Louly, and C. Prins. A reactive GRASP and path relinking for a combined production–distribution problem. *Computers and Operations Research*, 34(11): 3402–3419, 2007.

Cetinkaya, S., H. Uster, G. Easwaran, and B. B. Keskin. An integrated outbound logistics model for Frito-Lay: Coordinating aggregate-level production and distribution decisions. *Interfaces*, 39(5): 460–475, 2009.

Chandra, P. *On Coordination of Production and Distribution Decisions*. PhD dissertation, Department of Decision Sciences, The Wharton School, University of Pennsylvania, Philadelphia, 1989.

Chandra, P. and M. L. Fisher. Coordination of production and distribution planning. *European Journal of Operational Research*, 72(3): 503–517, 1994.

Chen, W.-H. and J.-M. Thizy. Analysis of relaxations for the multi-item capacitated lot-sizing problem. *Annals of Operations Research*, 26(1): 29–72, 1990.

Chen, C. L. and W. C. Lee. Multi-objective optimization of multi-echelon supply chain networks with uncertain product demands and prices. *Computers and Chemical Engineering*, 28(6–7): 1131–1144, 2004.

Chopra, S. and P. Meindl. *Supply Chain Management: Strategy, Planning, and Operations*. Upper Saddle River, NJ: Prentice Hall, 2001.

Christofides, N., A. Mingozzi, and P. Toth. The vehicle routing problem. In N. Christofides, A. Mingozzi, P. Toth, and C. Sandi (eds.), *Combinatorial Optimization*. Chichester: John Wiley & Sons, 1979.

Clarke, G. and J. W. Wright. Scheduling of vehicles from a central depot to a number of delivery points. *Operations Research*, 12(4): 568–581, 1964.

Dhaenens-Flipo, C. and G. Finke. An integrated model for an industrial production distribution problem. *IIE Transactions*, 33(9): 705–715, 2001.

Dogan, K. and M. Goetschalckx. A primal decomposition method for the integrated design of multi-period production-distribution systems. *IIE Transactions*, 31(11): 1027–1036, 1999.

Eksioglu, S. D., B. Eksioglu, and H. E. Romeijn. A Lagrangean heuristic for integrated production and transportation planning problems in a dynamic, multi-item two-layer supply chain. *IIE Transactions*, 39(2): 191–201, 2007.

Erenguc, S. S., N. C. Simpson, and A. J. Vakharia. Integrated production/distribution planning in supply chains: An invited review. *European Journal of Operational Research*, 115(2): 219–236, 1999.

Fahimnia, B., L. Luong, and R. Marian. An integrated model for the optimisation of a two-echelon supply network. *Journal of Achievements in Materials and Manufacturing Engineering*, 31(2): 477–484, 2008a.

Fahimnia, B., L. Luong, and R. Marian. Optimization/simulation modeling of the integrated production-distribution plan: An innovative survey. *WSEAS Transactions on Business and Economics*, 5(3): 52–65, 2008b.

Fahimnia, B., R. Z. Farahani, R. Marian, and L. Luong. A review and critique on integrated production-distribution models and techniques. *Journal of Manufacturing Systems*, 32(1): 1–19, 2013.

Fumero, F. and C. Vercellis. Integrating distribution, machine assignment and lot-sizing via Lagrangean Relaxation. *International Journal of Production Economics*, 49(1): 45–54, 1997.

Fumero, F. and C. Vercellis. Synchronized development of production, inventory, and distribution schedules. *Transportation Science*, 33(3): 330–340, 1999.

Gebennini, E., R. Gamberini, and R. Manzini. An integrated production–distribution model for the dynamic location and allocation problem with safety stock optimization. *International Journal of Production Economics*, 122(1): 286–304, 2009.

Geismar, H. N., G. Laporte, L. Lei, and C. Sriskandarajah. The integrated production and transportation scheduling problem for a product with a short lifespan. *INFORMS Journal on Computing*, 20(1): 21–33, 2008.

Gen, M. S. and A. Syarif. Hybrid genetic algorithm for multi-time period production/distribution planning. *Computers and Industrial Engineering*, 48(4): 799–809, 2005.

Gillett, B. and L. Miller. A heuristic algorithm for the vehicle dispatch problem. *Operations Research*, 22(2): 340–349, 1974.

Gilmore, P. and R. Gomory. Sequencing a one-state variable machine: A solvable case of the traveling salesman problem. *Operations Research*, 12(5): 675–679, 1964.

Grötschel, M. and M. W. Padberg. On the symmetric travelling salesman problem I: Inequalities. *Mathematical Programming*, 16(1): 265–280, 1979.

Jang, Y.-J., S.-Y. Jang, B.-M. Chang, and J. Park. A combined model of network design and production/distribution planning for a supply network. *Computers and Industrial Engineering*, 43(1–2): 263–281, 2002.

Kannan, G., P. Sasikumar, and K. Devika. A genetic algorithm approach for solving a closed loop supply chain model: A case of battery recycling. *Applied Mathematical Modeling*, 34(3): 655–670, 2010.

Karakitsiou, A. and A. Migdalas. A decentralized coordination mechanism for integrated production–transportation–inventory problem in the supply chain using Lagrangian relaxation. *Operational Research*, 8(3): 257–278, 2008.

Lee, Y. H. and S. H. Kim. Production-distribution planning in supply chain considering capacity constraints. *Computers and Industrial Engineering*, 43(1–2): 169–190, 2002.

Lee, Y. H., S. H. Kim, and C. Moon. Production-distribution planning in supply chain using a hybrid approach. *Production Planning and Control*, 13(1): 35–46, 2002.

Lei, L., S. Liu, A. Ruszczynski, and S. Park. On the integrated production, inventory, and distribution routing problem. *IIE Transactions*, 38(11): 955–970, 2006.

Lei, L., H. Zhong, and W. A. Chaovalitwongse. On the integrated production and distribution problem with bidirectional flows. *INFORMS Journal on Computing*, 21(4): 585–598, 2009.

Lejeune, M. A. and A. Ruszczynski. An efficient trajectory method for probabilistic production–inventory–distribution problems. *Operations Research*, 55(2): 378–394, 2007.

Liang, T. F. and H. W. Cheng. Application of fuzzy sets to manufacturing/distribution planning decisions with multi-product and multi-time period in supply chains. *Expert Systems with Applications*, 36(2): 3367–3377, 2009.

Liu, S., L. Lei, and S. Park. On the multi-product packing-delivery problem with fixed route. *Transportation Research Part E-Logistics and Transportation Review*, 44(3): 350–360, 2008.

Mula, J., D. Peidro, M. Diaz-Madronero, and E. Vicens. Mathematical programming models for supply chain production and transport planning. *European Journal of Operational Research*, 204(3): 377–390, 2010.

Naso, D., M. Surico, B. Turchiano, and U. Kaymak. Genetic algorithms for supply-chain scheduling: A case study in the distribution of ready-mixed concrete. *European Journal of Operational Research*, 177(3): 2069–2099, 2007.

Park, Y. B. An integrated approach for production and distribution planning in supply chain management. *International Journal of Production Research*, 43(6): 1205–1224, 2005.

Rizk, N., A. Martel, and S. D'Amours. Multi-item dynamic production–distribution planning in process industries with divergent finishing stages. *Computers and Operations Research*, 33(12): 3600–3623, 2006.

Rosencrantz, D. J., R. E. Steams, and P. M. Lewis. Approximate algorithms for the TSP. *Proceedings 15th IEEE Symposium on Switching and Automata Theory*, 33–42, 1974.

Ruokokoski, M., O. Solyali, J.-F. Cordeau, R. Jans, and H. Sural. Efficient formulations and a branch-and-cut algorithm for a production-routing problem. *GERAD Technical Report G-2010-66*, HEC Montreal, Canada, 2010.

Sabri, E. H. and B. M. Beamon. A multi-objective approach to simultaneous strategic and operational planning in supply chain design. *Omega*, 28(5): 581–598, 2000.

Safaei, A. S., S. M. M. Husseini, R. Z. Farahani, F. Jolai, and S. H. Ghodsypour. Integrated multi-site production-distribution planning in supply chain by hybrid modeling. *International Journal of Production Research*, 48(14): 4043–4069, 2010.

Sarmiento, A. M. and R. Nagi. A review of integrated analysis of production distribution systems. *IIE Transactions*, 31(11): 1061–1074, 1999.

Shen, Z. M. and L. Qi. Incorporating inventory and routing costs in strategic location models. *European Journal of Operational Research*, 179(2): 372–389, 2007.

Thomas, D. J. and P. M. Griffin. Coordinated supply chain management. *European Journal of Operational Research*, 94(1): 1–15, 1996.

Wang, H. Y., D.-C. Liu, T. Xing, and L. Zheng. A dynamic model for serial supply chain with periodic delivery policy. *International Journal of Production Research*, 48(3): 821–834, 2010.

Yilmaz, P. and B. Catay. Strategic level three-stage production distribution planning with capacity expansion. *Computers and Industrial Engineering*, 51(4): 609–620, 2006.

Yimer, A. D. and K. Demirli. A genetic approach to two-phase optimization of dynamic supply chain scheduling. *Computers and Industrial Engineering*, 58(3): 411–422, 2010.

Yossiri, A., J.-F. Cordeau, and R. Jans. Optimization-based adaptive large neighborhood search for the production routing problem. *Transportation Science*, doi:10.1287/trsc.1120.0443, 2012.

Yung, K.-L., J. Tang, A. W. H. Ip, and D. Wang, Heuristics for joint decisions in production, transportation, and order quantity. *Transportation Science*, 40(1): 99–116, 2006.

4

Increasing the Resiliency of Local Supply Chain Distribution Networks against Multiple Hazards

Sarah G. Nurre, Thomas C. Sharkey, and John E. Mitchell

CONTENTS

ABSTRACT We examine the resiliency of retail locations of a supply chain network to aid in the recovery of the local community after an extreme event. A two-stage stochastic programming model is used to determine the placement of permanent generators at the retail locations of Stewart's Shops, which distributes both convenience items and fuel in Upstate New York and Vermont, to enhance the resiliency of the supply chain. Our measure of resiliency specifically considers the recovery process of each retail location after the extreme event and its interdependency on other external infrastructure systems. Our computational experiments consider the multiple distinct types of hazards that can affect the retail locations of Stewart's Shops. We empirically explore different stochastic sampling procedures to solve the resiliency model. The results of computational tests indicate that we can converge to an approximate optimal solution quickly. We compare the resiliency efforts when planning for different types of hazards versus all hazards simultaneously as well as the impact of external infrastructure systems on the resiliency efforts. The empirical study identifies that the stores in rural, less densely populated areas serving a large population should be selected to receive generators.

KEY WORDS: *multiple hazards, resilience, supply chain network, two-stage stochastic program.*

4.1 Introduction

Recent extreme events, such as Hurricanes Irene in 2011 and Sandy in 2012 that affected New York and New Jersey, have demonstrated the need for enhancing the resiliency of supply chain systems. This is especially important for local supply chain networks that move critical goods, such as food, batteries, and fuel, into the areas affected by the extreme event. These critical goods allow the local population to begin to recover from the event and, often, companies operate hybrid retail operations that are part convenience stores (to provide food and batteries) and part gas station (to provide fuel). As a motivating example, Stewart's Shops is a company that operates 330 convenience stores and gas stations locations in Upstate New York and Vermont. Figure 4.1 presents a map of the retail locations of this company relative

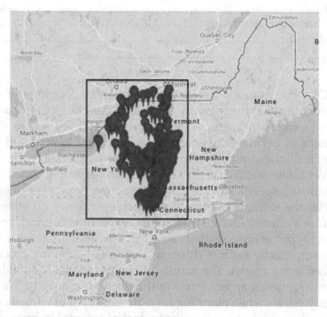

FIGURE 4.1
Locations of Stewart's Shops stores.

to the northeastern United States. Figures in the remainder of this chapter zoom in to the black box for clarity. In 2013, Stewart's had 58 locations that operated only as convenience stores and 272 locations that operated as both convenience stores and gas stations (see Stewart's Shops, 2013). In the past few years, multiple types of hazards have impacted Stewart's retail locations including hurricanes (in particular, Hurricanes Irene and Sandy), flooding, blizzards, and ice storms. This means that any efforts to increase the resiliency of Stewart's Shops for delivering critical goods to local populations should incorporate the potential impact of these various hazards.

The resiliency of a local supply chain distribution network, like that of Stewart's Shops, is typically focused on its ability to bounce back from disruptions. For local distribution networks, an important aspect of its bounce back is the capability to have its retail operations open for business. There are both internal and external factors that determine when a retail operation can begin its vital role as a distribution point after the disruptive event. The internal factors typically involve the steps necessary to reopen the store after any damage that was caused by the event. Example internal factors include (1) clearing debris, snow, or ice from the parking and refueling areas; (2) cleaning up the interior of the store and restocking shelves; (3) rebooting information systems; and (4) having workers arrive at the store, which often depends on external transportation systems. The external factors involve whether the services (such as power and telecommunications) necessary to support the retail operations are available after the event. For example, after Hurricane Sandy, lack of electrical power was a major source of the delay in the reopening of gas stations in the New York/New Jersey areas (see, e.g., Ma, 2012; Goldberg, 2012; Hu and Yee, 2012; Lipton and Krauss, 2012; Zernike, 2012). In fact, generators were brought into the area by certain gasoline companies for the sole purpose of reopening their points of distribution (Goldberg, 2012). In addition, many gas stations had their telecommunications services disrupted by Hurricane Sandy, implying that they were able to accept only cash from customers (e.g., Hu and Yee, 2012). Therefore, our proposed resiliency models specifically incorporate the reliance of reopening convenience and fuel distribution on the services provided by other (potentially disrupted) infrastructure systems.

The focus of this work is on locating *permanent* generators at the retail operations of a local supply chain distribution network to increase the resiliency of the system against multiple types of hazards. The resiliency of a particular retail operation is measured as the sum of the weighted (by demand) opening time for various commodities (e.g., cash-paying customers for fuel). The calculation of the opening time of a commodity at a retail operation will incorporate both the internal and external factors (e.g., dependencies on power and telecommunications) that affect it. This means that we use a two-stage stochastic preplanning model in which the first stage decisions locate the generators and the second stage, for each scenario, captures the resiliency of the distribution network for a realization of the damage of a hazard.

We do note that the "opening times" will not need to incorporate the arrival of inventory into the retail operation—the types of items that tend to be sold immediately after a disruptive weather event will be well stocked before that event and are often able to be replenished shortly after the event. Therefore, the focus of the resiliency measure should be on the ability to distribute on-site inventory rather than receiving shipments from elsewhere. It should be noted that Stewart's Shops will send out more "necessity" items to stores in areas potentially affected by an incoming event. In addition, the lack of power at gas stations was a major concern with respect to the gasoline shortages and rationing after Hurricane Sandy. Therefore, the particular focus on opening times of retail operations is well justified. This implies that we do not necessarily need to model the underlying warehouse–retailer distribution network in each scenario, thus allowing us to develop fast algorithms to solve our resulting resiliency models.

Our work is related to research on planning for supply chain disruptions; Snyder et al. (2006) and Snyder (2006) provide an overview of design and fortification models in this mean. These two-stage stochastic models they describe consider the location of supply chain components as the first-stage decisions and customer assignment as the second-stage decisions. The model in this chapter is distinct from this work because our first stage changes the properties of existing components and our second stage focuses on the time to recovery of distribution points.

There has been previous research on two-stage preplanning models for locating emergency supplies before a disaster. These models incorporate the dependencies of the supply chain on the transportation network by having travel times be scenario-dependent based on the damage of the event—see, for example, Shen et al. (2009), Mete and Zabinsky (2010), Rawls and Turnquist (2010), and Salmeron and Apte (2010), and Van Hentenryck et al. (2010). However, these models *do not* consider the reliance of the supply chain on other critical civil infrastructure systems, such as power and telecommunications. Shen (2013) examined building in new arcs in a network to increase the resiliency of interdependent infrastructure systems; however, these models assume that the infrastructures will "work together" in the second stage in terms of planning their recovery efforts from the event. This is often not the case for infrastructure restoration and, more importantly, local supply chain distribution networks will tend not to have a voice in restoration efforts from large-scale events. Further, these previous works on preplanning models tend to focus on scenarios that are generated from a single type of event, such as an earthquake (Dodo et al., 2007; Liu et al., 2009), rather than multiple hazards.

Other two-stage preplanning models examine the resilience of infrastructure systems subject to extreme events. Liu et al. (2009) examined the retrofitting of transportation networks; Miller-Hooks et al. (2012) examined a freight transportation network and determine the optimal allocation of preparedness and recovery actions; and Peeta et al. (2010) considered a highway network and sought to maximize the connectivity after the event. All of these studies do not incorporate the impact of external factors, such as power outages, on the recovery or resilience of the transportation network. The time it takes to bounce back to normal after an extreme event is one dimension of resiliency, which is often ignored, in these previous works that is specifically included in our model. A notable exception is Sheu (2007), who examined the time until relief is distributed after a disaster. However, Sheu (2007) focused on a multiobjective model in which the supply chain is not explicitly impacted but instead the demand is dynamic based on the degree of impact to different geographical locations.

The main contributions of this work include the following: (1) the consideration of multiple types of hazards in supply chain resiliency planning; (2) development of a two-stage stochastic program to enhance the resiliency of local supply chain networks specifically considering the recovery process of each retail operation and its interdependency on external infrastructures; and (3) an empirical exploration of stochastic sampling procedures to solve resiliency models.

The work proceeds as follows. Section 4.2 introduces the mathematical model and associated algorithm used to solve the two-stage stochastic supply chain resiliency problem; Section 4.3 presents the results of the computational analysis including the inclusion of different distribution of hazards and the impact of internal and external factors; and we conclude in Section 4.4. Please see the Appendix for corresponding parameter and data generation for each hazard scenario.

4.2 Mathematical Model and Algorithms

The proposed mathematical model to increase the resiliency of local supply chain distribution networks involves locating generators at retail operations to minimize the weighted "opening time" of the retail operations across a set of scenarios. The *opening time* of the retail operation for commodity l at store j requires that all internal and external factors that could prevent the opening are complete. In particular, the internal factors include (1) completing all work (such as repairing damage) for tasks that do not necessarily require power (we refer to these as "nonpower tasks") and (2) completing all work on tasks that do require power (we refer to these as "power-based tasks"). Note that there may be nonpower tasks (such as cleaning the store) that can be completed faster if power is available at the store—this will be incorporated in our model. The external factors include (1) any (potential) flooding around the store subsiding, (2) power being restored to the store (if there is no generator), and (3) telecommunications being restored to the store (for credit-only customers).

To provide a formal description of the model, it is necessary to provide an overview of the notation used to describe various parameters associated with the problem. The relevant parameters include the following:

- S is the set of all scenarios.
- N is the set of all stores.
- L is the set of all commodities.
- w_s is the weight of scenario $s \in S$. This would typically be viewed as the probability of the event occurring; however, for our purposes, this weight reflects the fact that we care about multiple types of hazards. A scenario for a type of hazard would have a probability, and the importance of that hazard multiplied by this probability would give us the weight for the scenario.
- d_{jl} is the cash demand for store j and commodity l.
- $\bar{d}_{jl}$ is the credit-only demand for store j and commodity l.
- K is the number of generators available.
- r_{sj}^p is the release time (or restoration time) for power (from the grid) at store j in scenario s.
- r_{sj}^c is the release for communications at store j in scenario s.
- r_{sj}^f is the release time for any floodwaters at store j in scenario s. We assume that $r_{sj}^f \leq r_{sj}^p$ for all s and j because power will not be restored to areas damaged by flooding until the flooding subsides and proper electrical inspections are done (see, e.g., Issler and Brodsky, [2012] for a discussion of this issue after Hurricane Sandy).
- p_{sjl} is the time needed to complete power-based tasks at store j for commodity l in scenario s.
- w_{sjl} is the work needed to be completed at store j for commodity l and scenario s for nonpower tasks.
- σ_j^p is the speed the work associated with nonpower tasks is completed when power is available.
- σ_j^{np} is the speed the work associated with non-power tasks is completed when power is not available.

The mathematical model is a two-stage stochastic program in which the first stage decisions locate generators at the retail operations of the local distribution supply chain network and the second stage decisions calculate the opening times of each store and commodity in each scenario. To this end, we define binary decision variables z_j for $j \in N$ that represent the decision of locating a generator at store j. The decision variables C_{sjl} and $\bar{C}_{sjl}$ provide the opening time of commodity l at store j in scenario s for cash customers and credit-only customers, respectively. The mathematical model of our resiliency model for local supply chain distribution networks (R-LSC) is then

$$\min_{z,C,\bar{C}} \sum_{s \in S} w_s \sum_{j \in N} \sum_{l \in L} d_{jl} C_{sjl} + \bar{d}_{sjl} \bar{C}_{sjl} \tag{4.1}$$

subject to

$$\sum_{j \in N} z_j \leq K \tag{4.2}$$

$$C_{sjl} \geq r_{sj}^p (1 - z_j) + p_{sjl} \quad \forall s \in S, \forall j \in N, \forall l \in L \tag{4.3}$$

$$C_{sjl} \geq \left(r_{sj}^p + \frac{w_{sjl} - \left(r_{sj}^p - r_{sj}^f \right) \sigma_j^{np}}{\sigma_j^p} \right)(1 - z_j) + \overset{*}{p}_{sjl} \quad \forall s \in S, \forall j \in N, \forall l \in L \tag{4.4}$$

$$C_{sjl} \geq \left(r_{sj}^f + \frac{w_{sjl}}{\sigma_j^p} \right) z_j + p_{sjl} \quad \forall s \in S, \forall j \in N, \forall l \in L \tag{4.5}$$

$$\bar{C}_{sjl} \geq C_{sjl} \quad \forall s \in S, \forall j \in N, \forall l \in L \tag{4.6}$$

$$\bar{C}_{sjl} \geq r_{sj}^c \quad \forall s \in S, \forall j \in N, \forall l \in L \tag{4.7}$$

$$z_j \in \{0, 1\} \quad \forall j \in N. \tag{4.8}$$

The constraint in Equation 4.2 limits the number of generators placed at the stores while the constraints in Equations 4.3 through 4.7 help to ensure the opening times consider all internal and external factors. In particular, for cash customers, the constraints in Equations 4.3 through 4.5 ensure that the following steps need to be complete prior to opening store j for commodity l: (1) flooding at the store subsides, (2) all nonpower tasks are completed, (3) power returns (either through a generator or being restored) to the store, and (4) all power-based tasks are completed. If $z_j = 0$, then the constraints in Equation 4.3 imply that we do not begin the power-based tasks until *at least* power is restored to the store and the constraints in Equation 4.4 imply that we do not begin the power-based tasks until *at least* the nonpower tasks are complete. The first term inside the parentheses in the constraints in Equation 4.4 is the time when we begin processing nonpower tasks at their "power" speed while the second term provides the amount of time required to finish the remaining work on these tasks. Note that if the tasks can be completed before the power speed kicks on, then the second term will be negative, and thus the constraints in Equation 4.3 will be active. The constraints in Equations 4.6 and 4.7 ensure the opening time for commodity l of credit-only customers is based on the fact that the store is open for cash customers for commodity l and communications is restored to the store.

It can be expected that the number of scenarios in R-LSC will be extremely large, and therefore it may not be computationally feasible to solve the form of R-LSC that incorporates all scenarios. Therefore, we will apply sample average approximation (see Shapiro et al., 2009) to determine an (approximate) optimal solution to R-LSC. For our case study of applying R-LSC to locating generators at Stewart's Shops, the set of scenarios span four distinct types of hazards: hurricanes, flooding, blizzards, and ice storms. The weight of a particular scenario, w_s, will be based on (1) the priority of the type of hazard associated with scenario s and (2) the probability that this type of hazard produces scenario s. For example, if we care about all four hazards equally, and a particular hurricane scenario has a 0.5 probability of occurring and a particular flooding scenario has a 0.2 probability of occurring, then the weights will be selected as 0.5 and 0.2 for these scenarios, respectively.

Our computational analysis will first apply sample average approximation (SAA) techniques to R-LSC to locate generators to increase the resiliency against a single type of hazard. The purpose of this is twofold: First, it will help determine the limits of planning for a particular type of hazard versus planning for all hazards and second, it will be the basis for an empirical sampling mechanism when protecting against multiple types of hazards. We will then explore different ways to sample when considering the multiple types of hazards and see their impact on the convergence of the SAA.

It turns out that R-LSC can be solved in $O(|N||S||L|)$ time. The method to solve R-LSC comes from the observation that the second-stage decisions are decomposable by store and that the generator location decisions can be viewed as *improving* the worst case opening times for each scenario. In particular, we define R_{sj} to be the "base" resiliency measure, that is, the resiliency of store j in scenario s when a generator is not located at j:

$$R_{sj} = \sum_{l \in L} d_{jl} \left(\max \left\{ r^p_{sj}, r^p_{sj} + \frac{w_{sjl} - \left(r^p_{sj} - r^f_{sj} \right) \sigma^{np}_j}{\sigma^p_j} \right\} + p_{sjl} \right)$$

$$+ \bar{d}_{jl} \left(\max \left\{ \max \left\{ r^p_{sj}, r^p_{sj} + \frac{w_{sji} - \left(r^p_{sj} - r^f_{sj} \right) \sigma^{np}_j}{\sigma^p_j} \right\} + p_{sjl}, r^c_{sj} \right\} \right)$$

This calculation simply forces the constraints in Equations 4.3 and 4.4 and 4.6 and 4.7 to be "active" for scenario s and store j. We then define R'_{sj} to be the objective for scenario s at store j when a generator is located there:

$$R'_{sj} = \sum_{l \in L} d_{jl} \left(r^f_{sj} + \frac{w_{sjl}}{\sigma^p_j} + p_{sjl} \right) + \bar{d}_{jl} \left(\max \left\{ r^f_{sj} + \frac{w_{sjl}}{\sigma^p_j} + p_{sjl}, r^c_{sj} \right\} \right).$$

Note that $R_{sj} \geq R'_{sj}$ for all s and j because $r^f_{sj} \leq r^p_{sj}$. We can then reformulate R-LSC as

$$\min \sum_{s \in S} w_s \sum_{j \in N} R_{sj}(1-z_j) + R'_{sj}z_j = \sum_{j \in N} \sum_{s \in S} w_s R_{sj} + \sum_{j \in N} \sum_{s \in S} w_s \left(R'_{sj} - R_{sj} \right) z_j \quad (4.9)$$

subject to

$$\sum_{j \in N} z_j \leq K \qquad (4.10)$$

$$z_j \in \{0, 1\} \quad \forall j \in N \qquad (4.11)$$

This reformulation is a knapsack problem in which all items (stores) have unit coefficients in the knapsack constraint. Therefore, the K stores with the lowest values of $\sum_{s \in S} w_s \left(R'_{sj} - R_{sj} \right)$ will be chosen to have generators located at them. It requires $O(|S||L|)$ time to calculate R_{sj} and R'_{sj} for each store, and therefore R-LSC can be solved in $O(|N||S||L|)$.

4.3 Computational Analysis

The purpose of this section is to explore both empirical convergence properties of R-LSC and to provide policy-based analysis for Stewart's Shops. The area in which Stewart's Shops operate their retail stores is prone to four types of hazards: hurricanes, flooding, blizzards, and ice storms. Each of these hazards has its own unique properties in terms of how it comes into the area and what damage it tends to cause to the retail stores, power grid, and telecommunications infrastructure. The Appendix discusses our techniques to generate a scenario for each of these distinct types of hazards.

For our case study, we consider two types of commodities: the "convenience" (e.g., food, water, and batteries) commodity and the "fuel" commodity. The demand level for these commodities at store $j \in N$ are a function of the location of the store, its surrounding areas, and its capabilities. We first determine the overall demand level for commodity $l \in \{c, f\}$ (where c = convenience and f = fuel) through the following procedure: (1) for each county and commodity l, determine the set of Stewart's Shops in that county capable of delivering that commodity and (2) split the demand (which we measure as the population; see NYS Data Center, Census, [2010] of the county evenly among all stores capable of delivering that commodity. We then assume that 50% of this overall demand for store j and commodity l is "credit-only" to determine d_{jl} and $\bar{d}_{jl}$. As an example, if there is a county with 10,000 residents and 10 Stewart's Shops that have gas stations, then we assume that the overall demand for fuel at each of these locations is 1000. Figure 4.2 displays a heat

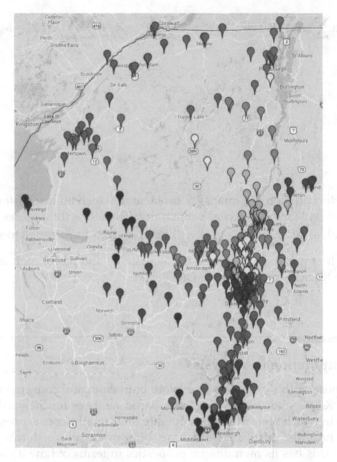

FIGURE 4.2
Heat map displaying the 330 Stewart's Shops stores and their respective total demand level, where darker shading indicates a higher level of demand.

map of the Stewart's stores based on their total level of demand over all types of customers (cash and credit-only) and commodities (convenience and fuel).

There are a few limitations to creating our demand levels in this manner. First, it assumes that the entire population will visit a Stewart's Shops store after an event. However, this assumption is not too limiting because if Stewart's market share is uniform across all counties, the optimal solution will not change because all solutions will have their objective function multiplied by their percentage market share. The second assumption is that it does not factor in the "closest" Stewart's Shops store to a given population—there may be a store across the street that is in a different county, and therefore the demand is assigned elsewhere. However, the number of these situations across all 330 stores is probably quite small. Further, the assignment of this demand assumes that the population will visit only one Stewart's Shops store and will not visit another one if the store is closed because of the event, as we assume static and not scenario-based demand. This assumption should be relaxed in future work; however, we justify the assumption by concluding a "non Stewart's Shop" will often be closer to a closed Stewart's Shops store than another Stewart's location and, therefore, the customers for the closed store will visit the "non Stewart's Shop."

4.3.1 Sample Average Approximation for Single Hazard Resiliency

Our first set of tests seeks to determine the location of generators among the 330 Stewart's Shops when considering each type of hazard individually. Specifically, we determine 12 solutions, one for each of the four hazards and three generator levels, $K = 17, 33$, and 50 which represent locating generators at 5%, 10%, and 15% of the number of stores, respectively. As with any empirical stochastic programming problem with a large number of scenarios, we must determine the appropriate number of scenarios to consider that provides a close approximation of the true optimal objective function value and solution. Our approach and stopping criteria uses a combination of the implementations presented by Linderoth et al. (2006) and Kleywegt et al. (2002).

For a fixed number of scenarios $|S|$, we solve M instances of the problem over $|S|$ randomly generated scenarios and determine their associated solutions $z_1,...z_M$ and objective function values $v_1,...v_M$. We then solve one instance of the problem by including the union of all $|S|$ scenarios over the M iterations for a total of $M|S|$ scenarios. Denote the solution and objective function value of this instance as z_{M+1} and v_{M+1}. We then calculate the optimality gaps of $v_1,...v_M$ relative to v_{M+1}. If all of the optimality gaps are within $\pm 1\%$, we stop; otherwise we increase the number of scenarios considered. The specifics of the algorithm are outlined in Algorithm 4.1.

Algorithm 4.1 Convergence Stopping Criteria

1. Set Boolean variable *stopping_criteria* equal to 0
2. Set number of scenarios, $|S| = 500$
3. Input: Number of generators, K
4. Input: Number of iterations, M
5. **while** *stopping_criteria* = 0 **do**
6.   **for** $i = 1:M$ **do**
7.   Generate $M|S|$ independent scenarios
8.   Solve for objective function value v_i and solution z_i for locating K generators using the $M|S|$ scenarios
9.   **end for**
10.   Consider all $M|S|$ scenarios generated
11.   Solve for the objective function value v_{M+1} and solution z_{M+1} for locating K generators using the $M|S|$ scenarios

12.   Calculate optimality gaps, $\frac{v_{M+1} - v_i}{v_{M+1}}$, for each $i = 1:M$

13.   if All optimality gaps are within ±1% then
14.   Set *stopping_criteria* = 1
15.   else
16.   Set number of scenarios, $|S| = |S| + 500$
17.   end if
18. end while
19. Return v_{M+1} and z_{M+1}

Because we are considering each type of hazard individually, each scenario has the same weight, specifically $\frac{1}{|S|}$. Figure 4.3 displays the results of these tests by showing the types of hazards, the three numbers of generators

Hazard	Number of generators (K)	Number of scenarios for convergence	Objective function value
Hurricane	0	–	12,042,400
	17	10,500	11,241,600
	33	11,500	10,876,600
	50	10,000	10,550,900
Flood	0	–	8,129,770
	17	37,500	7,284,080
	33	34,000	6,936,210
	50	29,000	6,886,730
Blizzard	0	–	9,543,590
	17	3,500	8,629,480
	33	3,500	8,237,490
	50	5,500	7,873,700
Ice storm	0	–	12,696,300
	17	7,000	11,270,400
	33	5,500	10,473,200
	50	8,000	9,784,940

FIGURE 4.3
Single hazard computational results.

considered, the resulting number of scenarios needed to converge, and objective function value. For reference, we include the objective function value when no generators are placed. From Figure 4.3, we see that the flooding hazard requires the greatest number of scenarios to converge. This is due to the fact that only 75 stores can ever be impacted by a flood, and further, more than two thirds of the stores are only impacted by at most one particular flooding event. This means that the intersection of stores impacted by two distinct flooding events is small, making it hard to decide on where to locate generators unless many scenarios are considered. Also from the table, we note that the hurricanes and ice storms have the biggest impact for reopening locations as is realized through the higher objective function values.

Figures 4.4 through 4.7 display the solutions for the three generator levels and their associated hazard where darker markers indicate the Stewart's Shops selected to receive a generator. For all four types of hazards, the solutions are incremental as we increase the number of available generators. In other words, the solution to the problem with 17 generators is a subset of the solution with 33 generators, and the solution to the problem with 33 generators is a subset of the solution with 50 generators. This is expected for a fixed set of scenarios as a direct result of the knapsack formulation. If for each store the R_{sj} and R'_{sj} values sufficiently converge, we can rank the objective function coefficient values and select the best K locations to receive generators for any number of generators K.

Therefore, the fact that our solutions exhibit this quality validates the use of our convergence stopping criteria. In practice, this solution property is desirable because if Stewart's Shops wanted to add more generators to their set of stores, they would not have to relocate existing generators to attain the optimal solution for the increased number. Instead they could determine

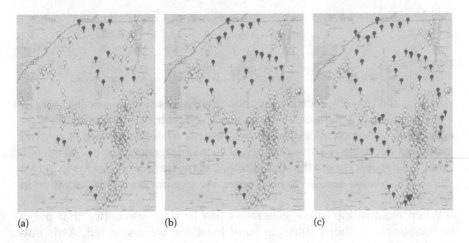

(a)   (b)   (c)

FIGURE 4.4
Hurricane hazard solutions. (a) 17 generators, (b) 33 generators, and (c) 50 generators.

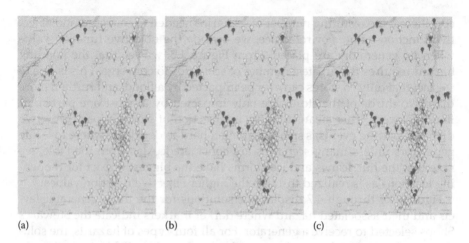

FIGURE 4.5
Flood hazard solutions. (a) 17 generators, (b) 33 generators, and (c) 50 generators.

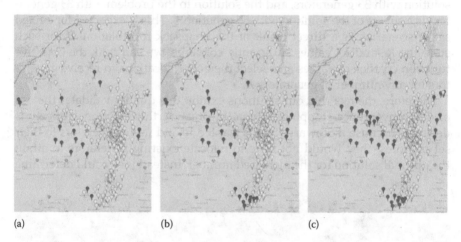

FIGURE 4.6
Blizzard hazard solutions. (a) 17 generators, (b) 33 generators, and (c) 50 generators.

which of the existing locations without generators will be selected for the installation of a generator. Also, as a direct result of the formulation, we notice the phenomenon of diminishing returns where the benefit of an extra generator decreases as the number of available generators is considered. From the figures, we see that the solutions to the hurricane, blizzard, and ice storm hazards are similar to each other. We expand on this observation in Section 4.3.3.

When examining where generators are located, we notice that many of the Stewart's Shops stores in rural locations were selected. As is outlined in the Appendix, the population density in the area surrounding a store impacts the time when power and communications are restored to

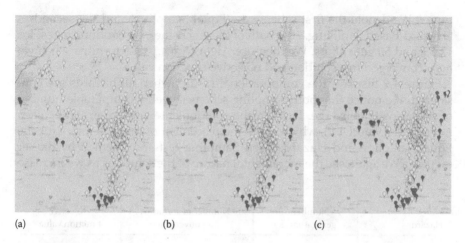

(a)   (b)   (c)

FIGURE 4.7
Ice storm hazard solutions. (a) 17 generators, (b) 33 generators, and (c) 50 generators.

the store, where power to urban locations is restored more quickly than to rural locations. Therefore, we see that rural locations serving a large population that have to wait longer for external services are often selected to receive generators.

4.3.2 Sample Average Approximation for All Hazard Types

We now seek to determine a solution that considers all four types of hazards. With these tests, we consider two different scenario sampling and weighting schematics. The first scheme looks at each type of hazard equally by incorporating an equal number of scenarios of each hazard type and distributing the weight equally across all scenarios. Specifically if $|S|$ scenarios are considered, we generate $\frac{|S|}{4}$ hurricane scenarios, $\frac{|S|}{4}$ flood scenarios, $\frac{|S|}{4}$ blizzard scenarios, and $\frac{|S|}{4}$ ice storm scenarios all with a weight equal to $\frac{|S|}{4}$. The second scheme considers each hazard equally; however, it incorporates the results from Section 4.3.1 by generating different number of scenarios for each hazard type. Let s_i^k denote the number of scenarios needed to converge to a solution for hazard type i and generator level k (e.g., set hurricane as hazard type 1, then $s_1^{17} = 10,500$ as seen in Figure 4.3). Under second sampling scheme, the number of scenarios generated for each type is $\frac{s_i^k |S|}{s_1^k + s_2^k + s_3^k + s_4^k}$.

The weight is then assigned by hazard type, where the sum of the weights for all scenarios of a set hazard type equals 0.25. The weight for each scenario of hazard type i (assuming four hazard types) equals $\frac{0.25\left(s_1^k + s_2^k + s_3^k + s_4^k\right)}{s_i^k |S|}$, which is equivalent to 0.25 divided by the number of scenarios of hazard type *i*. We denote the first scheme as "Equal Sampling" and the second as "Hazard-Based Sampling." For both of these schemes, we continue to test for 17, 33, and 50 generators. We then use Algorithm 4.1 to determine when we have converged to a solution and objective function value.

Figure 4.8 displays the computational results when all hazards are considered. We notice that both sampling schemes need a comparable number of scenarios to converge. Further, the solutions are similar as can be seen in Figures 4.9 and 4.10, which show the placement of the generators for the Equal Sampling and Hazard-Based Sampling tests represented by darker markers. We again see the incremental nature of the solutions as we increase

Hazard	Number of generators (K)	Number of scenarios for convergence	Objective function value
Equal sampling	17	12,000	9,806,580
	33	12,500	9,382,530
	50	8,000	8,994,290
Hazard-based sampling	17	12,000	9,820,900
	33	14,000	9,426,040
	50	9,000	8,905,650

FIGURE 4.8
All hazards computational results.

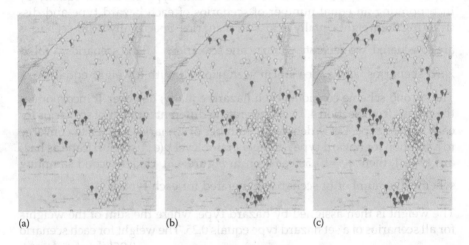

(a)   (b)   (c)

FIGURE 4.9
Solutions for all hazards with equal sampling. (a) 17 generators, (b) 33 generators, and (c) 50 generators.

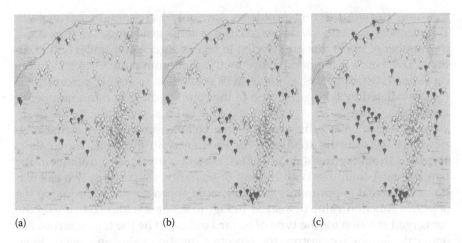

(a)   (b)   (c)

FIGURE 4.10
Solutions for all hazards with hazard-based sampling. (a) 17 generators, (b) 33 generators, and (c) 50 generators.

the number of generators considered. For both Equal Sampling and Hazard-Based Sampling the solution with 17 generators is a subset of the solutions with 33 and 50 generators and the solution with 33 generators is a subset of the solution with 50 generators. We reinforce that this is a nice solution property, as if budgets increase to allow for the installation of more generators, the optimal solution will not require that existing generators be relocated. We explore solution similarity by examining the solutions when hazards are considered individually and collectively.

4.3.3 Comparison of Solutions

We now examine the solutions, that is, where we selected to locate the generators, across the different types of hazards individually and collectively. We compare these solutions using two different metrics.

The first metric evaluates the converged solution to one test instance under s_i^k randomly selected scenarios of another test instance for hazard type i. For example, we take the approximate optimal solution when only hurricane scenarios are considered with 17 generators and evaluate it under a test instance with 37,500 flood scenarios (from Figure 4.3) and 17 generators. We perform this pairwise comparison for each of the hazards individually (hurricane, flood, blizzard, and ice storm) and under the two different sampling techniques (Equal Sampling and Hazard-Based Sampling). An optimality gap is then calculated as the percentage difference between the evaluated solution's objective function value and the optimal objective function value for that set of scenarios. The second metric that we use to compare solutions is a solution matching percentage which is calculated as the ratio of matching selected locations to the number of generators. For example, we take the solutions to two different problems, say, only hurricane scenarios with 17 generators and only blizzard scenarios with 33 generators. We then count the number of stores that are selected to receive a generator under both solutions. This count is bounded above by the minimum number of generators considered, which in this example is 17. A solution matching percentage is then calculated by taking the ratio of the count to the minimum number of generators.

Figure 4.11 displays the results of the comparisons using the first metric where we evaluate solutions. For each entry in the table, we create s_i^k scenarios (from Figures 4.3 and 4.8) consistent with the descriptions in the two left most columns. With this set of scenarios we perform two calculations: (1) solve it to optimality (which should approximately be equal to the objective function value displayed in Figure 4.3 or 4.8) and (2) evaluate the converged solution for the type of hazard indicated by the topmost row. For both calculations we capture the objective function value and calculate an optimality gap by taking the percentage difference between the evaluated solutions objective function value and the optimal objective function, which is displayed. From these results, it appears the flood hazard is least consistent with the other types of hazards as is represented by large optimality gaps. Further, it appears the Equal Sampling solution (column) performs better than the Hazard-Based Sampling solution (column) when evaluated against the different test instances as it almost always has a smaller optimality gap.

The results of the comparison using the second metric calculating a solution matching percentage are presented in Figures 4.12 and 4.13. The values are presented in Figure 4.12 and a corresponding heat map where darker values signify closer to 1 (100%) are presented in Figure 4.13. The calculations create a symmetric matrix; however, for conciseness we leave the values below the diagonal empty. First, we note that our previous observation about the incremental nature of the solutions is verified as all values within the same hazard–hazard comparison equal 1.00 (e.g., hurricane 17 and hurricane 33). It appears that flood hazards are least similar to ice storm, blizzard, and hurricane hazards. An important point is that the placement of generators change significantly when we move from a single hazard to all hazards. Therefore, it is important for local supply chain distribution networks to understand their goals for their resiliency efforts and incorporate the appropriate types of hazards into their analysis. Lastly, we point out that the solutions for Equal Sampling and Hazard-Based Sampling have a very high matching percentage, indicating that the biased sampling is not critical when calculating our resiliency efforts.

4.3.4 Impact of Internal and External Factors

In the last set of tests, we examine how internal and external factors impact where we locate generators. The opening time of a store depends on work

Hazard	K	Hurricane	Flood	Blizzard	Ice storm	Equal sampling	Hazard-based
Hurricane	17	–	4.98%	2.11%	2.69%	2.68%	4.86%
	33	–	7.15%	3.16%	3.87%	2.99%	4.13%
	50	–	9.63%	3.53%	4.41%	3.19%	4.88%
Flood	17	9.92%	–	11.31%	11.14%	4.85%	6.88%
	33	12.09%	–	13.64%	15.18%	5.91%	8.47%
	50		–	13.37%	13.64%	6.43%	11.63%
Blizzard	17	3.85%	8.63%	–	1.92%	2.39%	5.66%
	33	5.5%	11.84%	–	1.60%	2.58%	8.35%
	50	5.56%	16.01%	–	2.07%	2.51%	7.46%
Ice storm	17	5.88%	9.09%	2.22%	–	2.75%	5.07%
	33	9.50%	13.48%	2.22%	–	3.26%	6.29%
	50	9.78%	19.20%	2.53%	–	3.62%	6.88%
Equal sampling	17	3.09%	3.96%	1.75%	1.60%	–	1.84%
	33	4.01%	5.78%	2.28%	1.96%	–	2.19%
	50	3.39%	9.03%	1.91%	2.14%	–	2.26%
Hazard-based sampling	17	5.57%	8.77%	2.64%	0.88%	2.13%	–
	33	9.74%	14.32%	2.44%	1.49%	1.88%	–
	50	9.65%	21.75%	3.17%	1.77%	2.17%	–

FIGURE 4.11

Optimality gaps when the converged solution to the top row hazard is evaluated under the scenarios and optimal solution to the hazard described in the first column.

Hazard	K	Hurricane			Flood			Blizzard			Ice storm			Equal sampling			Hazard-based		
		17	33	50	17	33	50	17	33	50	17	33	50	17	33	50	17	33	50
Hurricane	17	1.00	1.00	1.00	0.12	0.12	0.12	0.41	0.41	0.41	0.29	0.29	0.29	0.29	0.41	0.53	0.29	0.41	0.53
	33		1.00	1.00	0.24	0.18	0.18	0.59	0.33	0.36	0.29	0.24	0.24	0.35	0.39	0.48	0.41	0.39	0.48
	50			1.00	0.59	0.42	0.28	0.76	0.58	0.50	0.53	0.52	0.40	0.65	0.73	0.62	0.71	0.73	0.62
Flood	17				1.00	1.00	1.00	0.06	0.12	0.41	0.00	0.12	0.35	0.59	0.76	0.76	0.53	0.76	0.82
	33					1.00	1.00	0.06	0.09	0.27	0.00	0.09	0.24	0.59	0.42	0.55	0.53	0.42	0.58
	50						1.00	0.06	0.09	0.18	0.00	0.09	0.18	0.59	0.42	0.36	0.53	0.42	0.38
Blizzard	17							1.00	1.00	1.00	0.41	0.76	0.88	0.47	0.76	0.94	0.53	0.76	0.94
	33								1.00	1.00	0.88	0.79	0.85	0.53	0.58	0.94	0.59	0.58	0.94
	50									1.00	1.00	0.88	0.76	0.82	0.79	0.80	0.82	0.79	0.80
Ice storm	17										1.00	1.00	1.00	0.41	0.82	1.00	0.41	0.82	1.00
	33											1.00	1.00	0.53	0.61	0.88	0.59	0.61	0.91
	50												1.00	0.76	0.73	0.70	0.76	0.73	0.70
Equal sampling	17													1.00	1.00	1.00	0.88	1.00	1.00
	33														1.00	1.00	1.00	0.97	1.00
	50															1.00	1.00	1.00	0.96
Hazard-based	17																1.00	1.00	1.00
	33																	1.00	1.00
	50																		1.00

FIGURE 4.12
Percentage of matching solutions using the count metric.

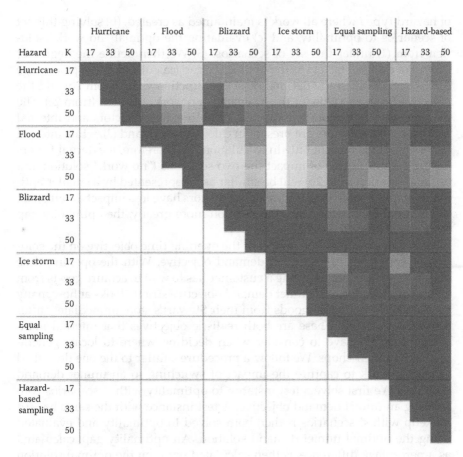

FIGURE 4.13
Heat map corresponding to the values shown in Figure 4.12 where darker represents closer to 1.

being completed at the store (including tasks that require and do not require power) and the restoration of external services, such as power, communications, and transportation (e.g., flooding subsiding). We first examine the degree of impact the external factors have on the opening time of each store. To quantify this degree of impact, we perform the following procedures for each hazard and generator level. We define the opening time for a store under a "no work" situation as the time when power and transportation is restored (for cash customers) and power, transportation, and communications is restored (for credit-only customers). (1) We first solve for the optimal "no work" solution by (a) generating s_i^k random scenarios (from Figures 4.3 and 4.8) of hazard type i where all work is removed and (b) solving this sets of scenarios to optimality. (2) We seek to evaluate the optimal "no work" solution under scenarios with work by (a) generating another set of s_i^k scenarios of hazard type i where all work is maintained as created, (b) solving this set of scenarios to optimality, and (c) evaluating the optimal "no work" solution from (1) under this set of scenarios. (3) We then define the degree of impact of external factors as the optimality gap (calculated as the percentage difference) between the optimal objective function value (from 2b) and the objective function value for the evaluated "no work" solution (from 2c). The placement of generators is impacted by both external factors and internal factors. External factors are present in calculations (1) and (2), while internal factors (work) is present only in calculation (2). Therefore, if external factors have a greater degree of impact the two solutions ("no work" solution and optimal solution from 2b) will be similar as is represented by a smaller optimality gap (closer to 0%). If the external factors have less impact and instead the internal factors influence the solution more greatly, the optimality gap from calculation (3) will be greater.

Second, we examine the impact of the opening time objective on the solution as compared to an unmet demand objective. With the opening time objective, we capture how long a customer has to wait to acquire goods from a Stewart's location. An unmet demand objective strictly looks at how many customers cannot attain goods from their Stewart's store immediately after a hazardous event. These are both realistic objectives that internal management would have to consider when deciding where to locate generators among their shops. We follow a procedure similar to the one described removing work to capture the impact of switching to an unmet demand objective. We first solve a test instance to optimality with s_i^k scenarios considering an unmet demand objective. A test instance with the same hazard makeup with s_i^k scenarios is then both solved to optimality and evaluated using the optimal unmet demand solution. An optimality gap, calculated as a percentage difference, is then calculated between the optimal solution objective value and evaluated solution objective value. An optimality gap closer to 0% emphasizes that the two objectives are interchangeable.

The results of the two sets of tests are presented in Figure 4.14. The column labeled External Factors captures the degree of impact each instance has on external factors, where closer to 0% is a greater dependence on external factors. We see that the flood hazard has the greatest dependence on external factors. The result makes sense, as the flooding hazard is the only hazard that depends on the release times for flooding, power, and communications. The column labeled Unmet Demand captures the interchangeability of the opening time and unmet demand objectives, where closer to 0% represents a higher degree of interchangeability. On first inspection, the optimality gaps appear not close to zero, thereby signifying that the two objectives differ. However, Figure 4.15 displays the 330 Stewart's Shops locations and the count of times each location was selected to have a generator across all possible test scenarios (opening time, opening time with no work, and unmet demand), where a darker color represents a higher count. When we compare Figure 4.15

Hazard	K	External factors	Unmet demand
Hurricane	17	2.10%	3.09%
	33	2.06%	4.72%
	50	2.19%	6.95%
Flood	17	0.64%	5.86%
	33	0.49%	6.24%
	50	0.23%	2.68%
Blizzard	17	1.66%	2.54%
	33	1.60%	2.69%
	50	1.73%	4.34%
Ice storm	17	1.13%	1.80%
	33	1.37%	1.79%
	50	1.30%	5.46%
Equal sampling	17	1.52%	2.51%
	33	1.68%	2.84%
	50	1.45%	5.18%
Hazard-based sampling	17	1.66%	2.50%
	33	1.33%	3.11%
	50	1.50%	4.80%

FIGURE 4.14
Optimality gaps capturing the interdependence on external services and interchangeability of different objectives.

to Figure 4.2 we will see that the higher demand stores correspond to those stores with a higher selection count to receive a generator. This indicates that demand is a strong driver in the selection of generator placement under any objective. However, note that many high-demand stores (particularly those near the Albany, New York area in the center right of the map) are not often selected to receive a generator. These high-demand stores are in areas with highly dense populations that often have power restored quickly. It is the high-demand stores in rural, less densely populated areas, that are selected to receive generators more frequently, as these stores traditionally have to wait longer for restoration of power and communications.

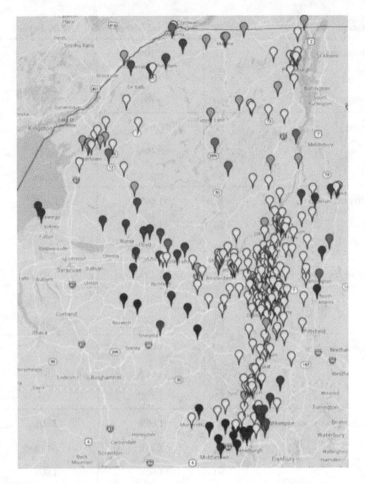

FIGURE 4.15
Heat map displaying the 330 Stewart's Shops stores and their respective count of instances where they are selected to receive a generator. A darker color indicates a higher count.

4.4 Conclusions

We examined the resiliency of retail locations of a supply chain network to aid in the recovery of the local community after an extreme event. A two-stage stochastic program was used to determine the location of permanent generators among Stewart's Shops 330 stores. Consistent with recent events, we considered four types of hazardous situations that could impact Stewart's Shops: hurricanes, flooding, blizzards, and ice storms. Using an optimal greedy algorithm, we tested the model for a variety of different instances by considering each hazard individually and collectively using two different sampling procedures. We examined the impact of external factors, such as power, communications, and flooding, and different objectives on the solution obtained. The results demonstrate that we are able to empirically converge to an optimal solution, using a hybrid stopping criteria, by considering a relatively small number of scenarios. We observed that the solutions were incremental as the number of available generators increased, which is a direct result of the knapsack formulation. This is a desirable property as a prioritized list of stores that should receive generators can be made and followed as more generators become available.

Future work should involve sensitivity analysis on the parameters used to generate the hours of work that each store is required to perform both with and without power. These parameters impact the opening times of stores and ultimately if demand can be met. Further, scenario-dependent demand should be incorporated into the model. Currently, we consider only static demand but for many areas, there are many Stewart's Shops close to one another that can service customers if their preferred shop is closed.

Disclaimer

The views expressed in this chapter are those of the authors and do not reflect the official policy or position of the United States Air Force, Department of Defense, or the United States Government.

Appendix: Scenario Generation

In this section, we describe the procedures to generate a scenario for each type of hazard. For each of the four hazards (hurricane, flood, blizzard, and ice storm), we describe the different parameters that factor into the scenario generation. There are various *continuous* parameters that impact the scenario, meaning that it is not possible to generate the probability of a particular scenario based on these parameters. For each type of hazard, the parameters to consider for a scenario are

- *Hazard characteristics.* Track and intensity level (hurricane, blizzard, ice storm) or body of water (flood). Figure 4.16 provides the starting and ending points for hurricanes, blizzards, and ice storms. Hurricanes travel south to north and blizzards and ice storms travel west to east.

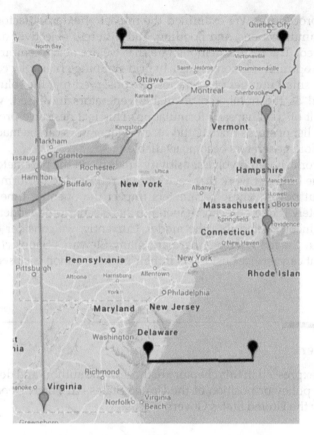

FIGURE 4.16
Potential starting and ending points for hurricane, blizzard, and ice storm tracks.

- Probability of an impact for each store $j \in N$ along with whether the store is impacted for that scenario
- Release times for power, communications, and flooding (i.e., r_{sj}^p, r_{sj}^c, and r_{sj}^f) for each impacted store $j \in N$
- Amount of work associated with nonpower tasks (i.e., w_{sjl}) for each impacted store $j \in N$ and commodity $l \in L\$$
- Speed with which nonpower tasks are processed both with $\left(\sigma_j^p\right)$ and without $\left(\sigma_j^{np}\right)$ power available
- Amount of time needed to process power-based tasks (p_{sjl})

We now present the values and logic behind each of these parameters for the different types of hazards.

Hurricane

- *Hazard characteristics.* We randomly sample a starting (on the south horizontal line in Figure 4.16) and ending point (on the north horizontal line in Figure 4.16) for the track and assume the track of the storm is a straight line between these points. We then calculate the distance from this track to each store, which we represent as δ_{sj}. Each storm has an associated intensity, $\alpha \in \{0, 1, 2\}$ where 0 represents a tropical storm, 1 represents a category 1 hurricane, and 2 represents a category 2 hurricane.

- *Probability of impact for each store.* The probability of a store being impacted by the storm track used in scenario s is calculated as follows:

$$p_{sj} = \begin{cases} 0.95\left(0.2+\dfrac{\alpha}{4}\right) & \text{if } 0 \le \delta_{sj} < 5 \text{ miles} \\[2mm] 0.90\left(0.2+\dfrac{\alpha}{4}\right) & \text{if } 5 \le \delta_{sj} < 10 \text{ miles} \\[2mm] 0.85\left(0.2+\dfrac{\alpha}{4}\right) & \text{if } 10 \le \delta_{sj} < 15 \text{ miles} \\[2mm] 0.80\left(0.2+\dfrac{\alpha}{4}\right) & \text{if } 15 \le \delta_{sj} < 20 \text{ miles} \\[2mm] 0.75\left(0.2+\dfrac{\alpha}{4}\right) & \text{if } 20 \le \delta_{sj} < 25 \text{ miles} \\[2mm] 0.70\left(0.2+\dfrac{\alpha}{4}\right) & \text{if } 25 \le \delta_{sj} < 30 \text{ miles} \\[2mm] 0.65\left(0.2+\dfrac{\alpha}{4}\right) & \text{if } 30 \le \delta_{sj} < 35 \text{ miles} \\[2mm] 0.60\left(0.2+\dfrac{\alpha}{4}\right) & \text{if } 35 \le \delta_{sj} < 40 \text{ miles} \\[2mm] 0.55\left(0.2+\dfrac{\alpha}{4}\right) & \text{if } 40 \le \delta_{sj} < 45 \text{ miles} \\[2mm] 0.50\left(0.2+\dfrac{\alpha}{4}\right) & \text{if } 45 \le \delta_{sj} < 50 \text{ miles} \\[2mm] 0.01\left(0.2+\dfrac{\alpha}{4}\right) & \text{if } 50 \le \delta_{sj} \end{cases} \qquad (4.12)$$

A random number, $v_{sj} \in [0, 1]$, is generated for each store j and scenario s. If $v_{sj} \leq p_{sj}$, then we classify store j as impacted by a power outage, otherwise not. If $v_{sj} \leq \dfrac{p_{sj}}{3}$, we classify store j as impacted by a communications outage, otherwise not. This means that we assume that every store impacted by a communications outage is also impacted by a power outage. This is consistent with the observations after Hurricane Sandy: Communications outages are typically either caused by a power outage to a local central office (implying power needs to be restored to the area) or because downed poles in the area carried both power lines and communications lines.

- *Release times for power, communications, and flooding.* For hurricanes, we will only consider release times for power and communications because our focus for this hazard is on wind damage (flooding from a hurricane is considered indirectly in the "flooding hazard"). Let ξ_j denote the population density of the county of store j. The release time for power is then calculated using the following equation:

$$r_{sj}^p = 2 + \tanh\left(\frac{\delta_{sj}}{4 \cdot \xi_j}\right) \cdot 70 \qquad (4.13)$$

which puts all power release dates in the range of $[2, 72]$ hours if the store is impacted and 0 otherwise.

- The release time of communications is then calculated as a function of the release time of power and whether there is a precedence constraint between repairing power and communications. This can happen in situations when the power company owns the poles that carry telecommunications lines. If there is a power precedence, then the release time of communications is greater than the release time of power. We specifically calculate the release of communications as

$$r_{sj}^c = \begin{cases} 0.75 r_{sj}^p & \text{if no power precedence over communications} \\ 1.25 r_{sj}^p & \text{if power precedence over communications} \end{cases}$$

where 50% of the stores are selected at random to have a power precedence.

- *Work associated with nonpower tasks.* For each impacted store j, we set the work for the convenience commodity to 6 hours and the work for the fuel commodity to 3 hours.

- *Processing speed of nonpower tasks.* Work can be completed at a rate of 1 unit if there is no power and 1.5 units if there is power available (restored early or a generator).

- *Time needed for power-based tasks.* At each store j, we set the processing time for the fuel commodity to 2 hours. For the convenience commodity, we set the processing time to 3 hours. If a store is prone to flooding from a specific storm we add on a random integer in [0, 15] to the processing time for convenience. We denote a storm prone to flooding if the store is within 10 miles of the track and within 0.1 mile of a body of water for tropical storms, 0.2 mile of a body of water for category 1 hurricanes, and 0.3 mile of a body of water for category 2 hurricanes.

Flooding

- *Hazard characteristics.* There are six large bodies of water that could flood and impact various locations of Stewart's Shops: (1) Hudson, (2) Mohawk, (3) Lake Champlain, (4) Black River, (5) Vermont Rivers, and (6) St. Lawrence River. Smaller local rivers and creeks flowing off of these bodies of water are also included in our analysis. A flood is selected uniformly at random from these 6. An intensity level, α, from {0.1, 0.2, 0.3, 0.4, 0.5} is selected uniformly at random for each flood that represents how far away from the river flooding occurs.

- *Probability of impact for each store.* There is no probability associated with the impact to a store. The impact is determined solely by the flooding event and the distance from the store to the body of water. For each store j we know its distance to each of the six bodies of water, denote this δ_{sj} for the flooding event in scenario s. If store j is closer to the body of water than the flood intensity level ($\delta_{sj} \le \alpha_s$), then it is considered impacted by the event.

- *Release times for power, communications, and flooding.* The release time (in hours) for flooding is calculated using the table that follows, where the leftmost column represents the intensity of the flood (α_s), and the topmost row represents the distance a store is from the body of water associated with s (δ_{sj}).

r_{sj}^f	0.1	0.2	0.3	0.4	0.5
0.1	20				
0.2	36	16			
0.3	50	30	14		
0.4	62	42	26	12	
0.5	72	52	36	22	10

The release time of power at store j under scenario s is calculated using the following equation:

$$r_{sj}^p = \max\left\{5+r_{sj}^f, 5+\tanh\left(\frac{200\cdot\alpha}{\xi_j}\right)\cdot 150\cdot(\alpha+0.1)\right\} \tag{4.14}$$

which puts the release dates in the range [15, 95] if a store is impacted and 0 otherwise.

The release time of communications is calculated using the same equation as was done for the hurricane hazard, specifically:

$$r_{sj}^c = \begin{cases} 0.75r_{sj}^p & \text{if no power precedence over communications} \\ 1.25r_{sj}^p & \text{if power precedence over communications} \end{cases}$$

- *Work associated with nonpower tasks.* For each impacted store j, we set the work for the convenience commodity to 12 hours and the work for the fuel commodity to 5 hours.
- *Processing speed of nonpower tasks.* Work can be completed at a rate of 1 unit if there is no power and 1.5 units if there is power available (restored early or a generator).
- *Time needed for power-based tasks.* The processing time for power-based tasks for the fuel commodity is set to 2 hours. The processing time for power-based tasks for the convenience commodity is calculated using the table below, where the leftmost column represents the intensity of the flood (α_s) and the topmost row represents the distance a store is from the body of water associated with s (δ_{sj}).

	0.1	0.2	0.3	0.4	0.5
0.1	10				
0.2	20	10			
0.3	30	20	10		
0.4	40	30	20	10	
0.5	50	40	30	20	10

Blizzard

- *Hazard characteristics.* We randomly sample a starting (on the west vertical line in Figure 4.16) and ending point (on the east vertical line in Figure 4.16) and assume the track of the storm is a straight line between these points. We then calculate the distance from this track to each store (denoted as δ_{sj}). Each storm also has an associated intensity, $\alpha \in \{1, 2, 3, 4, 5\}$ based on Northeast Snowfall Impact Scale.

- *Probability of impact for each store.* Let l_j denote the elevation of store j. The potential impact of the blizzard on the store is a function of its distance from the track, the intensity of the storm, and its elevation. The probability of a store being impacted by the storm track used in scenario s is calculated as follows:

$$
p_{sj} = \begin{cases}
0.95\left(0.2 + \dfrac{\alpha}{10} + \min\{0.3, l_j\}\right) & \text{if } 0 \le \delta_{sj} < 5 \text{ miles} \\[2mm]
0.90\left(0.2 + \dfrac{\alpha}{10} + \min\{0.3, l_j\}\right) & \text{if } 5 \le \delta_{sj} < 10 \text{ miles} \\[2mm]
0.85\left(0.2 + \dfrac{\alpha}{10} + \min\{0.3, l_j\}\right) & \text{if } 10 \le \delta_{sj} < 15 \text{ miles} \\[2mm]
0.80\left(0.2 + \dfrac{\alpha}{10} + \min\{0.3, l_j\}\right) & \text{if } 15 \le \delta_{sj} < 20 \text{ miles} \\[2mm]
0.75\left(0.2 + \dfrac{\alpha}{10} + \min\{0.3, l_j\}\right) & \text{if } 20 \le \delta_{sj} < 25 \text{ miles} \\[2mm]
0.70\left(0.2 + \dfrac{\alpha}{10} + \min\{0.3, l_j\}\right) & \text{if } 25 \le \delta_{sj} < 30 \text{ miles} \\[2mm]
0.65\left(0.2 + \dfrac{\alpha}{10} + \min\{0.3, l_j\}\right) & \text{if } 30 \le \delta_{sj} < 35 \text{ miles} \\[2mm]
0.60\left(0.2 + \dfrac{\alpha}{10} + \min\{0.3, l_j\}\right) & \text{if } 35 \le \delta_{sj} < 40 \text{ miles} \\[2mm]
0.55\left(0.2 + \dfrac{\alpha}{10} + \min\{0.3, l_j\}\right) & \text{if } 40 \le \delta_{sj} < 45 \text{ miles} \\[2mm]
0.50\left(0.2 + \dfrac{\alpha}{10} + \min\{0.3, l_j\}\right) & \text{if } 45 \le \delta_{sj} < 50 \text{ miles} \\[2mm]
0.01\left(0.2 + \dfrac{\alpha}{10} + \min\{0.3, l_j\}\right) & \text{if } 50 \le \delta_{sj}
\end{cases}
\tag{4.15}
$$

A random number, $v_{sj} \in [0, 1]$, is generated for each store j and scenario s. If $v_{sj} \leq p_{sj}$, then we classify store j as impacted by a power outage, otherwise not. If $v_{sj} \leq \frac{p_{sj}}{3}$, we classify store j as impacted by a communications outage, otherwise not.

- *Release times for power, communications, and flooding.* We consider only release times for power and communications for this type of hazard. Let ξ_j represent the population density in the area surrounding store j. The release time for power is calculated using the following equation:

$$r_{sj}^p = \alpha + \tanh\left(\frac{\delta_{sj}}{4 \cdot \xi_j}\right) \cdot 50 \qquad (4.16)$$

which puts all power release dates in the range of [1, 55] hours if the store is impacted and 0 otherwise.

The release of communications is calculated using the same equation as was done for the hurricane scenario, specifically:

$$r_{sj}^c = \begin{cases} 0.75 r_{sj}^p & \text{if no power precedence over communications} \\ 1.25 r_{sj}^p & \text{if power precedence over communications} \end{cases}$$

- *Work associated with nonpower tasks.* For each impacted store j we set the work for the convenience commodity to 4 hours and the work for the fuel commodity to 2 hours.

- *Processing speed of nonpower tasks.* Work can be completed at a rate of 1 unit if there is no power and 1.5 units if there is power available (restored early or a generator).

- *Time needed for power-based tasks.* For each impacted store j, we set the processing time for work requiring power for both the convenience commodity and the fuel commodity to 2 hours.

Ice Storm

- *Hazard characteristics.* We randomly sample a starting (on the west vertical line in Figure 4.16) and ending point (on the east vertical line in Figure 4.16) and assume the track of the storm is a straight line between these points. We then calculate the distance from this track to each store (denoted as δ_{sj}. Each storm also has an associated intensity, $\alpha \in \{1, 2, 3, 4, 5\}$ based on the Sperry–Piltz Ice Accumulation Index.

- *Probability of impact for each store.* Let l_j denote the elevation of store j. The potential impact of the blizzard on the store is a function of its distance from the track, the intensity of the storm, and its elevation. The probability of a store being impacted by the storm track used in scenario s is calculated as follows:

$$p_{sj} = \begin{cases} 0.95\left(0.2+\dfrac{\alpha}{10}+\min\{0.3,l_j\}\right) & \text{if } 0 \le \delta_{sj} < 5 \text{ miles} \\[2mm] 0.90\left(0.2+\dfrac{\alpha}{10}+\min\{0.3,l_j\}\right) & \text{if } 5 \le \delta_{sj} < 10 \text{ miles} \\[2mm] 0.85\left(0.2+\dfrac{\alpha}{10}+\min\{0.3,l_j\}\right) & \text{if } 10 \le \delta_{sj} < 15 \text{ miles} \\[2mm] 0.80\left(0.2+\dfrac{\alpha}{10}+\min\{0.3,l_j\}\right) & \text{if } 15 \le \delta_{sj} < 20 \text{ miles} \\[2mm] 0.75\left(0.2+\dfrac{\alpha}{10}+\min\{0.3,l_j\}\right) & \text{if } 20 \le \delta_{sj} < 25 \text{ miles} \\[2mm] 0.70\left(0.2+\dfrac{\alpha}{10}+\min\{0.3,l_j\}\right) & \text{if } 25 \le \delta_{sj} < 30 \text{ miles} \\[2mm] 0.65\left(0.2+\dfrac{\alpha}{10}+\min\{0.3,l_j\}\right) & \text{if } 30 \le \delta_{sj} < 35 \text{ miles} \\[2mm] 0.60\left(0.2+\dfrac{\alpha}{10}+\min\{0.3,l_j\}\right) & \text{if } 35 \le \delta_{sj} < 40 \text{ miles} \\[2mm] 0.55\left(0.2+\dfrac{\alpha}{10}+\min\{0.3,l_j\}\right) & \text{if } 40 \le \delta_{sj} < 45 \text{ miles} \\[2mm] 0.50\left(0.2+\dfrac{\alpha}{10}+\min\{0.3,l_j\}\right) & \text{if } 45 \le \delta_{sj} < 50 \text{ miles} \\[2mm] 0.01\left(0.2+\dfrac{\alpha}{10}+\min\{0.3,l_j\}\right) & \text{if } 50 \le \delta_{sj} \end{cases} \tag{4.17}$$

A random number, $v_{sj} \in [0, 1]$, is generated for each store j and scenario s. If $v_{sj} \le p_{sj}$, then we classify store j as impacted by a power outage, otherwise not. If $v_{sj} \le \dfrac{p_{sj}}{3}$, we classify store j as impacted by a communications outage, otherwise not.

- *Release times for power, communications, and flooding.* We consider only release times for power and communications. The release time for power is calculated using the following equation:

$$r_{sj} = \alpha + 1 + \tanh\left(\frac{\delta_{sj}}{4 \cdot \xi_j}\right) \cdot 80 \qquad (4.18)$$

which puts all power release dates in the range of [1, 86] hours if the store is impacted and 0 otherwise.

The release of communications is calculated using the same equation as was done for the hurricane scenario, specifically:

$$r_{sj}^c = \begin{cases} 0.75r_{sj}^p & \text{if no power precedence over communications} \\ 1.25r_{sj}^p & \text{if power precedence over communications} \end{cases}$$

- *Work associated with nonpower tasks.* For each impacted store j we set the work for the convenience commodity to 2 hours and the work for the fuel commodity to 1 hour.
- *Processing speed of nonpower tasks.* Work can be completed at a rate of 1 unit if there is no power and 1.5 units if there is power available (restored early or a generator).
- *Time needed for power-based tasks.* For each impacted store j, we set the processing time for work requiring power for both the convenience commodity and the fuel commodity to 2 hours.

Acknowledgment

Thomas C. Sharkey was supported in part by the National Science Foundation under grant number 1254258.

References

Dodo, A., R. Davidson, N. Xu, and L. Nozick. Application of regional earthquake mitigation optimization. *Computers & Operations Research*, 34(8): 2478–2494, 2007.

Goldberg, D. With one eye on recovery from Hurricane Sandy, state keeps watch on approaching nor'easter. *The Star Ledger*, November 4, 2012. Retrieved from http://www.nj.com/news/index.ssf/2012/11/with_one_eye_on_recovery_from .html (Accessed March 14, 2013).

Hu, W. and V. Yee. While fuel is promised, drivers wait hours for gas. The *New York Times*, November 3, 2012. Retrieved from http://www.nytimes.com/2012/11/05 /nyregion/while-fuel-is-promised-drivers-wait-hours-for-gas.html (Accessed February 7, 2013).

Issler, M. and R. Brodsky. Safety-check issues hamper power restoration. *Newsday*, November 10, 2012. Retrieved from http://www.newsday.com/long-island /safety-check-issues-hamper-power-restoration-1.4210515 (Accessed March 14, 2013).

Kleywegt, A. J., A. Shapiro, and T. Homem de mello. The sample average approxima-tion method for stochastic discrete optimization. *SIAM Journal on Optimization*, 12: 479–502, 2002.

Linderoth, J., A. Shapiro, and S. Wright. The empirical behavior of sampling meth-ods for stochastic programming. *Annals of Operations Research*, 142(1): 215–241, 2006.

Lipton, E. and C. Krauss. Military to deliver fuel to storm-ravaged region. The *New York Times*, November 2, 2012. Retrieved from http://www.nytimes .com/2012/11/03/business/military-to-deliver-fuel-to-storm-region.html (Accessed August 7, 2013).

Liu, C., Y. Fan, and F. Ordónez. A two-stage stochastic programming model for trans-portation network protection. *Computers & Operations Research*, 36(5): 1582–1590, 2009.

Ma, M. Power-hungry customers wait an hour for gas on Route 17 in Paramus. *NJ.com*, October 31, 2012. Retrieved from http://www.nj.com/bergen/index .ssf/2012/10/power-hungry_customers_wait_an_hour_for_gas_on_route_17 _in_paramus.html (Accessed August 15, 2013).

Mete, H. O. and Z. B. Zabinsky. Stochastic optimization of medical supply location and distribution in disaster management. *International Journal of Production Economics*, 126: 76–84, 2010.

Miller-Hooks, E., X. Zhang, and R. Faturechi. Measuring and maximizing resilience of freight transportation networks. *Computers & Operations Research*, 39(7): 1633–1643, 2012.

NYS Data Center Census 2010. Retrieved from http://esd.ny.gov/NYSDataCenter /Census2010.html (Accessed August 29, 2013).

Peeta, S., F. Salman, D. Gunnec, and K. Viswanath. Pre-disaster investment decisions for strengthening a highway network. *Computers & Operations Research*, 37(10): 1708–1719, 2010.

Rawls, C. G. and M. A. Turnquist. Pre-positioning of emergency supplies for disaster response. *Transportation Research Part B: Methodological*, 44(4): 521–534, 2010.

Salmeron, J. and A. Apte. Stochastic optimization for natural disaster asset preposi-tioning. *Production and Operations Management*, 19: 43–53, 2010.

Shapiro, A., D. Dentcheva, and A. Ruszczyński. *Lectures on Stochastic Programming: Modeling and Theory*. MPS/SIAM Series on Optimization, Vol. 9. Philadelphia: Society for Industrial and Applied Mathematics, 2009.

Shen, S. Two-stage models and algorithms for optimizing infrastructure design and recovery operations under stochastic disruptions. *Computers & Operations Research*, 40(11): 2677–2688, 2013.

Shen, Z., M. M. Dessouky, and F. Ordónez. A two-stage vehicle routing model for large-scale bioterrorism emergencies. *Networks*, 54(4): 255–269, 2009.

Sheu, J. An emergency logistics distribution approach for quick response to urgent relief demand in disasters. *Transportation Research Part E: Logistics and Transportation Review*, 43: 687–709, 2007.

Snyder, L. V. Facility location under uncertainty: A review. *IIE Transactions*, 38(7): 537–554, 2006.

Snyder, L. V., M. P. Scaparra, M. S. Daskin, and R. L. Church. Planning for disruptions in supply chain networks. In M. P. Johnson, B. Norman, and N. Secomandi (eds.), *TutORials 2006*, INFORMS Tutorials in O.R. Series, Chapter 9. Catonsville, MD: The Institute for Operations Research and the Management Sciences, 2006.

Stewart's Shops. Stewart's Shops locations. Retrieved from http://www.stewarts shops.com/find-a-shop/ (Accessed August 29, 2013).

Van Hentenryck, P., R. Bent, and C. Coffrin. Strategic planning for disaster recovery with stochastic last mile distribution. In *Integration of AI and OR Techniques in Constraint Programming for Combinatorial Optimization Problems* (pp. 318–333). Springer, Berlin, 2010.

Zernike, K. Gasoline runs short, adding woes to storm recovery. The *New York Times*, November 1, 2012. Retrieved from http://www.nytimes.com/2012/11/02/nyregion/gasoline-shortages-disrupting-recovery-from-hurricane.html (Accessed August 7, 2013).

5

Nested Partitions for Large-Scale Optimization in Supply Chain Management

Weiwei Chen and Leyuan Shi

CONTENTS

ABSTRACT In today's global business environment, an effective and reliable supply chain substantially increases a company's competitive advantage. Supply chain management has played a crucial role in a business's success. Optimization has been sought to model and solve many of these supply chain problems. However, as the size of the optimization model grows, the difficulty of finding the optimal or near-optimal solution typically grows exponentially. Nested partitions is one of the algorithms developed to tackle these large-scale optimization problems. In this chapter, the nested partitions method, and its implementation in two critical supply chain problems, are reviewed. One problem is the intermodal hub location problem, a special case of facility location problems, and the other is the multilevel capacitated lot-sizing problem with backlogging, a complex problem in production planning. The computational results show that the hybrid nested partitions approaches are superior to standard mathematical programming and specialized heuristic approaches, and thus provide an effective alternative in enhancing supply chain performance.

KEY WORDS: *facility location, large-scale optimization, metaheuristic, nested partitions, production planning, supply chain management, supply chain optimization.*

5.1 Introduction

In today's global market, designing and managing an effective and reliable supply chain increases a company's competitive advantage. There is no coincidence between the success of retailers such as Amazon and Walmart and their innovative supply chain strategies. To appreciate fully the importance of supply chain management, we first need to understand what a supply chain is.

5.1.1 Supply Chain and Supply Chain Management

A supply chain essentially touches all components of a business's value chain in fulfilling customer demands. It begins with purchasing raw materials to supply the production of the products or services that the company wants to sell. In making sourcing and procurement decisions, costs and quality are two major factors in consideration. Once the materials are in place, manufacturing of the product or operations to provide the service begins. The main objective is to lower the operations and inventory costs. Undoubtedly, production planning and scheduling is complex because of the various resources involved, such as machines, labors, and inventories. It further adds to the complexity when multiple operating facilities involve. Demand planning and forecasting is another crucial effort in supply chain operations. An accurate demand forecast not only ensures customer fulfillment and satisfaction, but also enables effective inventory management. Inventories may include final products, raw materials, and/or works-in-process. When the products are ready, the next element in the value chain is distribution of the products to retailers or customers. The speed of the distribution is critical in today's intense competition. Meanwhile, a company inevitably desires to lower the distribution costs. Hence, a good design of the distribution network remains at the top of key supply chain strategical decisions to be made. The effort may not end here, as customer services and reverse supply chain have become essential parts of the business. Reverse supply chain refers to the activities required to retrieve a used product from a customer, either by customer return or vendor recycling, and dispose of it or reuse it (Daniel and Wassenhove, 2002).

Although each of the aforementioned components of a supply chain are already difficult to manage, the ultimate benefit of supply chain management cannot be achieved unless parts of or all components are considered systematically and coordinately. For example, the design of a distribution network can be better realized when production planning as its upstream element and transportation as its downstream element are both taken into account. In like manner, production cannot be planned and scheduled accurately if demand forecast and procurement plans are not in place. With the global market and outsourcing, the situation becomes further complicated as the supply chains of involved companies are often coupled. It seems reasonable to design and operate a supply chain by considering others. However, it has seldom been achieved because of conflicting interests and objectives of various parties, not to mention the exponentially increasing technical difficulty to consider the entire system. For instance, suppliers typically appreciate large quantities of purchase from manufacturers so that they can maximize the profit and reduce the transportation cost. On the other hand, from manufacturers' viewpoint, a more flexible purchase plan can mitigate the risk of overproduction and reduce the inventory cost. Consequently, the changes of one company's supply chain unavoidably affect other companies' supply chains. Therefore, recent industrial trends pay more attention to the robustness of the supply chain as an effort to deal effectively with uncertainties and mitigate risks.

In short, a supply chain involves all business components and facilities, including suppliers, manufacturers, warehouses and distribution centers, and retailers, which make the product conform to customer requirements. The main objective of supply chain management is to reduce costs, improve efficiency, and mitigating risks, from both strategic and operational perspectives. We quote a definition of supply chain management from Simchi-Levi et al. (2008, p. 1) as follows:

> Supply chain management is a set of approaches utilized to efficiently integrate suppliers, manufacturers, warehouses, and stores, so that merchandise is produced and distributed at the right quantities, to the right locations, and at the right time, in order to minimize systemwide costs while satisfying service level requirements.

Such a definition perfectly introduces the main topic of this chapter: large-scale optimization in supply chain management. First, optimization is the modeling and solution techniques popularly used in supply chain management. As indicated in the definition, the objective is to minimize costs, while operational requirements and service level have to be satisfied as constraints. There exist other variants of the objective, such as to maximize profits, which can be translated into similar mathematical models. Second, the optimization models of interest are notably in large scale, as we strive to optimize from the system level and to coordinate different components in a supply chain. As the size of the optimization problem grows, the solution space grows exponentially, as well as the effort to identify the best or even a "good enough" solution. As a result, despite systemwide supply chain management being the ultimate goal, research efforts in supply chain optimization have been focusing on providing an optimal or near-optimal solution to one or multiple components of a supply chain.

5.1.2 Supply Chain Optimization

To survive in the market competition and compete for the narrow profit margin, supply chain optimization has become more and more popular among the world's leading retailers, manufacturers, and distributors. As mentioned in the preceding text, plenty of opportunities exist for supply chain optimization.

The battle typically begins at sourcing. Optimization can be used to design an effective sourcing strategy that maintains the production capability in a cost-saving manner. An example is the game-changing sourcing strategy deployed by Procter & Gamble (P&G; Sandholm et al., 2006). Suppliers make electronic offers that express rich forms of capabilities and efficiencies, while P&G, as a buyer, also expresses constraints and preferences. The sourcing solution brought these together via an optimization engine to determine the optimal allocation of business to the suppliers. It was reported to save P&G $294.8 million from $3 billion sourcing activities over a period of two and a half years.

Because of the problem complexity and importance, production planning and scheduling have drawn much attention in the optimization community. Although planning and scheduling are mentioned together here, they are concerned with activities in different time frames. Production planning is a medium-term activity, with the objective of allocating production plans to fulfill customer demands at minimum production and inventory costs. On the other hand, production operations scheduling is a short-term activity. The goal is to determine the optimal allocation of resources (e.g., machines and labors) and the sequencing and timing of tasks. Traditionally, production planning and operations scheduling have been studied independently in consideration of model tractability. Just to name a few, reviews of production planning models (e.g., lot-sizing models) and solution algorithms can be found in Drexl and Kimms (1997), Kreipl and Pinedo (2004), and Voß and Woodruff (2006), and reviews of operations scheduling include Pinedo and Chao (1999), Hall and Potts (2003), and Pinedo (2008). Some studies have also looked at the possible integration of planning and schedule from an optimization point of view, such as the work in Miller (2002) and Maravelias and Sung (2009). Not surprisingly, solving an integrated optimization model is much more difficult and typically requires a much longer computational time.

Another popular research area for optimization is distribution. Distribution refers to the steps taken to move and store a product from the supplier stage to a customer stage in the supply chain (Chopra, 2003). The problems can be further divided into strategic and operational decisions. A well-known strategic level decision is the distribution network design (Amiri, 2006). An effective distribution network delivers products in a timely manner, or strives for a low-cost solution, or ideally does both. The distribution network design directly affects the inventory costs, transportation costs, and facility setup and maintenance costs. Many researchers have extensively studied facility location problems to determine the optimal locations of warehouses and/or plants. As a strategic level decision making, the facility location decisions are usually targeted on one time change for long-term benefits, as the investments on each location can be millions of, or even billions of dollars. A variety of facility location models, as well as their integration with transportation or inventory, have been reviewed in Daskin et al. (2005) and Klose and Drexl (2005).

Given a distribution network, the mode of transportation and the operational schedule are then determined to transport the products from plants or warehouses to warehouses or retailers. Contrary to the network design, the operational transportation schedules are typically made on a daily basis and subject to uncertainties. From a transportation company standpoint, the optimization objective is usually to reduce the unnecessary costs incurred by moving empty tractors and/or trailers from one location to another, while the on-time delivery rate has to be maintained to fulfill customer services. Equivalently, the goal can also be to maximize the overall profit over a planning horizon. Variants of problems include vehicle routing, local pickup and delivery, and intermodal transportation. Typically, the load pickup and delivery window constraints and the driver capacity and working hour constraints pose difficulties to the mathematical models. Different formulations have been proposed and various algorithms have been developed, such as column generation (Xu et al., 2001), approximate dynamic programming (Powell et al., 2002), heuristics and data mining (Campbell and Savelsbergh, 2004; Bräsy and Gendreau, 2005a,b; Chen et al., 2013), and so forth.

In addition to the aforementioned topics, optimization has played an important role in other aspects of the supply chain management. Some notable examples include inventory management (Roy et al., 1997; Chen et al., 2007), reverse supply chain (Pishvaee et al., 2011), considering uncertainties in supply chain (Gupta and Maranas, 2003; Bertsimas and Thiele, 2004; Santoso et al., 2005), supply chain disruption (Arreola-Risa and DeCroix, 1998; Cui et al., 2010; Aboolian et al., 2013), and so forth. Not surprisingly, many of these optimization models are large scale and notoriously difficult to solve. In this chapter, we review the nested partitions algorithm, a metaheuristic optimization framework for solving large-scale optimization, in applications to some of these supply chain optimization problems.

The rest of this chapter is organized as follows. The nested partitions optimization framework is introduced in Section 5.2. In Sections 5.3 and 5.4, we review the implementation of nested partitions for solving facility location problems and production planning problems, respectively. Conclusions are drawn in Section 5.5.

5.2 Nested Partitions and Large-Scale Optimization

Many supply chain optimization problems, as well as other key business strategic and operational decisions, are large-scale discrete optimization problems. These problems are challenging and are notoriously difficult to solve.

5.2.1 Large-Scale Discrete Optimization

There are two major techniques for solving large-scale discrete optimization problems: (1) exact algorithms that are grounded in mathematical programming and dynamic programming theories and (2) heuristics, including metaheuristics, which aim to find acceptable solutions quickly.

Exact algorithms have been studied for decades. With the rapid growth of computational power and advancements in mathematical theories, breakthroughs in the ability to solve large-scale problems using mathematical programming have been achieved. Generally, branching methods and decomposition methods are two primary classes of mathematical programming methods used to solve discrete optimization problems (Wolsey, 1998; Nemhauser and Wolsey, 1999). Relaxation methods play a key role in the use of math programming for solving discrete optimization problems (Lemarechal, 2001; Fisher, 2004). Lagrangian relaxation can be thought of as a decomposition method with respect to the constraints because it moves one or more constraint into the objective function. The Lagrangian problem is easier to solve because the complicating constraints are no longer present. Furthermore, it often produces a fairly tight and hence useful bound. Dynamic programming is another class of exact algorithms that are popularly used to address sequential decision making. It solves subproblems recursively using the Bellman equation (Puterman, 2005). Despite the breakthroughs, it can still be very time consuming or structurally difficult when using exact algorithms for solving practical problems with large sizes. Some approximation algorithms have been developed based on similar mathematical theories. These algorithms guarantee that the solution lies within a certain range of the optimal solution, and usually provide provable runtime bounds (Hall and Hochbaum, 1986; Bertsimas and Teo, 1998).

Another class of optimization algorithms are heuristic algorithms, which aim to find good solutions within an acceptable time frame. Unlike exact algorithms, heuristics do not usually make a performance guarantee, such as bounds. But they have been proven to be very effective and efficient in solving many real-world problems. Some heuristic algorithms utilize the domain knowledge to speed up the search and thus are problem dependent. Meanwhile, there exist a class of metaheuristics, which are designed to be applicable to a large variety of problems. One simple algorithm is greedy search (Cormen et al., 1990), which makes the locally optimal decision at each step. It basically sacrifices solution quality for fast computational speed, whereas all heuristics have different trade-offs between these two key measurements. Tabu search (Cvijovíc and Klinowski, 1995; Glover and Laguna, 1997) improves local search by prohibiting the repeated visits to the same solution within a short period of time. A subcategory of the metaheuristics comprises evolutionary algorithms that are typically used by artificial intelligence and machine learning communities. These algorithms include genetic algorithms (Goldberg, 1989; Mitchell, 1998), ant colony optimization (Dorigo, 1992; Dorigo et al., 1999), particle swarm optimization (Kennedy and Eberhart, 1995; Poli et al., 2007), and so forth. Some metaheuristics use probability distributions as rules and strategies, such as simulated annealing (Kirkpatrick et al., 1983; van Laarhoven and Aarts, 1987), cross-entropy methods (Rubinstein and Kroese, 2004; De Boer et al., 2005), and model reference adaptive search (Hu et al., 2007). Metaheuristics are designed to be general enough to apply to many real problems, and it is usually easy to further speed up the algorithms by embedding domain knowledge.

5.2.2 Nested Partitions Overview

Introduced by Shi and Ólafsson (2000a), the nested partitions (NP) method is a metaheuristic framework for solving large-scale optimization problems. The NP method is a partitioning and sampling based strategy that focuses computational effort on the most promising region of the solution space while maintaining a global perspective on the problem. It is capable to be applied for solving deterministic (both discrete and continuous) and stochastic optimization, for example, simulation-based optimization (Shi and Ólafsson, 2000b) problems. The focus of this chapter is on deterministic discrete optimization. Consider the following optimization problem:

$$\min_{\theta \in \Theta} f(\theta) \tag{5.1}$$

Problem 5.1 can be a large-scale mixed integer program, or a combinatorial optimization problem. The feasible region is denoted as Θ, and an objective function $f : \Theta \rightarrow \Re$ is defined on this set. When the problem size is not large, efficient exact algorithms can usually be deployed to solve the problem with guaranteed optimal solutions. However, when the problem becomes large, no known exact algorithm exists today to solve the problem within reasonable time frame. This type of problems is what the NP method targets on.

In each iteration of the NP algorithm, we assume that there is a region (subset) of Θ that is considered the *most promising*. We partition this most promising region into a fixed number of M *subregions* and aggregate the entire *complementary region* (also called *surrounding region*) into one region, that is, all the feasible solutions that are not in the most promising region. Therefore we consider $M + 1$ subsets that are a partition of the feasible region Θ; namely, they are disjoint and their union is equal to Θ. This is referred to as a valid partitioning scheme. Each of these $M + 1$ regions is sampled using some random sampling scheme to generate feasible solutions that belong to that region. The performance values (objective values) of the randomly generated samples are used to calculate the *promising index* for each region. This index determines which region is the most promising region in the next iteration. If one of the subregions is found to be the best, this region becomes the next most promising region. The next most promising region is thus nested within the last. If the complementary region is found to be the best, then the algorithm *backtracks* to a larger region that contains the previous most promising region. This larger region becomes the new most promising region, and is then partitioned and sampled in the same fashion.

If region η is a subregion of region σ, we call σ a *superregion* of η. Let $\sigma(k)$ denote the most promising region in the kth iteration. We further denote the *depth* of $\sigma(k)$ as $d(k)$. The feasible region Θ has depth 0, the subregions of Θ have depth 1, and so forth. When Θ is finite, eventually there will be regions that contain only a single solution. Such singleton regions are called regions of *maximum depth*. If the problem is infinite, define the maximum depth to correspond to the smallest desired sets. The maximum depth is denoted as d^*. With this notation, the generic nested partitions algorithm is described in Algorithm 5.1 (Shi and Ólafsson, 2000a). The special cases of being at minimum or maximum depth are considered separately.

Algorithm 5.1 Nested Partitions Algorithm ($0 < d(k) < d^*$)

1. **Partitioning.** Partition the most promising region $\sigma(k)$ into M subregions $\sigma_1(k)$, ..., $\sigma_M(k)$, and aggregate the complementary region $\Theta \backslash \sigma(k)$ into one region $\sigma_{M+1}(k)$.

2. **Random sampling.** Randomly generate N_j sample solutions from each of the regions $\sigma_j(k)$, $j = 1, 2, \ldots, M + 1$:

$$\theta_1^j, \theta_2^j, \ldots, \theta_{N_j}^j, \; j = 1, 2, \ldots, M + 1.$$

Calculate the corresponding performance values:

$$f\left(\theta_1^i\right), f\left(\theta_2^i\right), ..., f\left(\theta_{N_j}^i\right), j = 1, 2, ..., M+1.$$

3. **Calculate promising index.** For each region σ_j, $j = 1, 2, ..., M + 1$, calculate the promising index as the best performance value within the region:

$$I(\sigma_j(k)) = \min_{i \in \{1, 2, ..., N_j\}} f\left(\theta_i^j\right), j = 1, 2, ..., M+1.$$

4. **Move.** Calculate the index of the region with the best performance value.

$$\hat{j}_k = \arg\min_{j \in \{1, ..., M+1\}} I(\sigma_j(k)).$$

If more than one region is equally promising, the tie can be broken arbitrarily. If this index corresponds to a region that is a subregion of $\sigma(k)$, that is $\hat{j}_k \le M$, then let this be the most promising region in the next iteration:

$$\sigma(k+1) = \sigma_{\hat{j}_k}(k).$$

Otherwise, if the index corresponds to the complementary region, that is $\hat{j}_k = M+1$, backtrack to the superregion of the current most promising region:

$$\sigma(k + 1) = \sigma(k - 1).$$

or backtrack to the entire solution space:

$$\sigma(k + 1) = \Theta.$$

For the special case of $d(k) = 0$, the steps are identical except there is no complementary region. The algorithm hence generates feasible sample solutions from the subregions and in the next iteration moves to the subregion with the best promising index. For the special case of $d(k) = d^*$, there are no subregions. The algorithm therefore generates feasible sample solutions from the complementary region and either backtracks or stays in the current most promising region.

The aforementioned procedure of the NP method gives us a framework that guides the search and enables convergence analysis (Shi and Ólafsson,

2000a). The basic idea is that the sequence of regions that NP visits, denoted by $\{\sigma(k)\}_{k=1}^{\infty}$, is a Markov chain, and the regions of maximum depth are the only absorbing states. That is, $\{\sigma(k)\}_{k=1}^{\infty}$ ends up in one of these absorbing states with probability 1.

Theorem 5.1

The NP method with a valid partitioning scheme for the optimization problem

$$\theta^* \in \arg \min_{\theta \in \Theta} f(\theta),$$

converges almost surely to a global minimum in finite time, that is, there exist a $K < \infty$ such that with probability 1,

$$\sigma(k) = \{\theta^*\}, \quad \forall k \geq K,$$

where

$$\theta^* \in \arg \min_{\theta \in \Theta} f(\theta). \qquad \blacksquare$$

The generic NP algorithm is particularly effective for problems where the solution space can be partitioned such that good solutions tend to be clustered together and the corresponding regions are hence natural candidates for concentrating the computational effort. For large-scale problems, it may not be natural to find such a partitioning scheme. Therefore, the real power of NP is usually achieved when incorporating domain knowledge, local search, or mathematical programming into its framework. The NP framework provides the guidance to explore the entire solution space with global convergence guarantee, while the incorporated algorithm exploits each subregion efficiently.

According to our computational experience, hybrid NP algorithms with local heuristics (including math programming) almost always outperform heuristics alone. It has been successfully used in solving optimization problems from various domains. In Shi et al. (2001), NP is combined with greedy search, dynamic programming, and genetic algorithm to solve a product design problem, which is to use the preferences of potential customers to design a new product such that the market share is maximized. Another example is a hybrid NP and tabu search algorithm developed to allocate optimally a given buffer capacity in a factory such that a desired system performance is achieved (Shi and Men, 2003). Other examples include radiation treatment planning (Zhang et al., 2009), feature selection in data mining (Ólafsson and Yang, 2005), and local pickup and delivery (Pi et al., 2008; Chen et al., 2013). In the next two sections, we present two such applications, one in facility location (Pi et al., 2008; Chen et al., 2009) and the other in production planning (Wu et al., 2011), and demonstrate how to develop hybrid NP methods for solving large-scale supply chain optimization.

5.3 Nested Partitions for Facility Location

In this section, we review a detailed implementation of the NP method for solving the intermodal hub location problem (IHLP), which is first published in Pi et al. (2008). IHLP is a hub location problem in the intermodal transportation industry and can be viewed as a special type of discrete facility location problems.

Truck/rail intermodal transportation combines the cost-effectiveness of rail with the flexibility of trucks. When freight is transported, it is first loaded onto a trailer or a container that is moved from the shipper to a nearby rail ramp by truck. Then the freight is conveyed by rail to a ramp close to its destination. Finally it is delivered to the consignee by truck (Chen et al., 2011). Intermodal operations utilize a sealed container or truck trailer that is mechanically moved between modes (truck, rail) in a seamless fashion. An intermodal terminal has equipments suitable for transferring the containers and trailers between modes. Because of its economic impact, this problem has drawn a great attention by researchers (Campbell, 1994; O'Kelly and Bryan, 1998; Racunica and Wynter, 2005).

5.3.1 Mathematical Model

The IHLP aims to minimize the total costs of the intermodal transportation system, including operations costs of the open hubs and the routing costs of the intermodal movements. The notations and formulation are introduced as follows:

Sets and indexes
- $I = \{1,\ldots,|I|\}$: set of terminal (origin and destination) locations
- $R = \{1,\ldots,|R|\}$: set of intermodal hub locations
- $F = \{1,\ldots,|F|\}$: set of demand flows, that is, product movements from an origin to a destination
- $O_f, f \in F$: origin terminal of flow f
- $D_f, f \in F$: destination terminal of flow f

Parameters
- $w_f, f \in F$: amount of flow f
- $c_r^O, r \in R$: operations cost of hub r, if r is open

- $c_{ij}^T, (i,j) \in I \times R \cup R \times R \cup R \times I$: transportation cost of unit flow between location i to location j
- $c_f^P, f \in F$: penalty cost per unit amount of flow f moved by other mode of transportation instead of intermodal, such as long-haul truck movement, which is usually more expensive
- $\delta_{ij}, (i,j) \in I \times R \cup R \times I$: equals 1 if a movement from i to j is allowed; equals 0 otherwise

Decision variables

- $x_{fkl}, f \in F, k \in R, l \in R$: the amount of flow f moved through intermodal rail line (k, l)
- $y_{ij}, (i,j) \in I \times R \cup R \times R \cup R \times I$: the amount of flow from location i to location j
- $z_r, r \in R$: binary facility location variables. $z_r = 1$ if hub r is open; $z_r = 0$ otherwise
- $u_f, f \in F$: the amount of flow f that is not moved through the intermodal operations

Formulation

min

$$\sum_{r \in R} c_r^O \cdot z_r \tag{5.2}$$

$$+ \sum_{(i,j)\in I\times R\cup R\times R\cup R\times I} c_{ij}^T \cdot y_{ij} \tag{5.3}$$

$$+ \sum_{f \in F} c_f^P \cdot u_f \tag{5.4}$$

Subject to

$$u_f + \sum_{k \in R, l \in R} x_{fkl} = w_f, \quad \forall f \in F, \tag{5.5}$$

$$\sum_{l \in R} x_{fkl} \leq w_f \cdot z_k, \quad \forall f \in F, k \in R, \tag{5.6}$$

$$\sum_{k \in R} x_{fkl} \leq w_f \cdot z_l, \quad \forall f \in F, l \in R, \tag{5.7}$$

$$\sum_{f\in F,\, l\in R:O_f=i} x_{fkl} \le y_{ik}, \quad \forall i\in I,\, k\in R, \tag{5.8}$$

$$\sum_{f\in F} x_{fkl} \le y_{kl}, \quad \forall k\in R,\, l\in R, \tag{5.9}$$

$$\sum_{f\in F,\, k\in I:D_f=j} x_{fkl} \le y_{lj}, \quad \forall l\in R,\, j\in I, \tag{5.10}$$

$$x_{fkl} \le w_f \cdot \delta_{O_f,k} \quad \forall f\in F,\, k\in I,\, l\in I, \tag{5.11}$$

$$x_{fkl} \le w_f \cdot \delta_{l,D_f} \quad \forall f\in F,\, k\in I,\, l\in I, \tag{5.12}$$

x_{fkl}, y_{ij}, u_f are nonnegative, and z_r are binary.

In the objective function, Equation 5.2 is the hub operations cost, Equation 5.3 is the intermodal transportation cost, and Equation 5.4 is the penalty of using another transportation mode instead of intermodal. The constraints in Equation 5.5 require that all the demand flows should be transported. The constraints in Equations 5.6 and 5.7 guarantee that flow can be routed only via an open hub in an intermodal movement. The constraints in Equations 5.8 through 5.10 are the flow balance requirements. The constraints in Equations 5.11 and 5.12 are the restrictions of the movements between terminals and hubs.

5.3.2 Hybrid NP and Mathematical Programming Algorithm

To effectively solve the IHLP, a hybrid NP and mathematical programming (HNP–MP) approach is developed. Math programming is used to solve the smaller subproblems.

In most NP implementations, complete sample solutions are generated in each sampling step. To be able to integrate the mixed integer programming (MIP) techniques into the NP framework, partial sampling is developed. A partial sample is a solutions that is generated by sampling only part of the variables in a given region. For example, assume that the current most promising region is $(\theta_1^*,\ldots,\theta_k^*, \theta_{k+1},\ldots,\theta_n)$, where $\theta_1^*,\ldots,\theta_k^*$ are fixed. By sampling $\theta_{k+1},\ldots,\theta_{k+1}(1 \le j < n-k)$, we have a partial sample in the following form: $(\theta_1^*,\ldots,\theta_k^*, \bar{\theta}_{k+1},\ldots,\bar{\theta}_{k+j}, \theta_{k+j+1},\ldots,\theta_n)$. The purpose of introducing partial samples is that by reducing the number of variables of the original problem, the subproblem with only variables $(\theta_{k+j+1},\ldots,\theta_n)$ could be solved effectively using math programming.

The HNP–MP algorithm is presented in Algorithm 5.2.

Algorithm 5.2 HNP–MP Algorithm

1. Set the initially most promising region as the entire solution space. Set the initial surrounding region as $\varnothing$.
2. If stopping conditions are met, stop and return the best sample obtained; otherwise, proceed to the next step.
3. Obtain LP solution for the current most promising region. Perform LP biased sampling (to be introduced later) in the most promising region and surrounding region to generate partial samples $(\theta_1^*,\ldots,\theta_k^*,\ \bar{\theta}_{k+1},\ldots,\bar{\theta}_{k+j},\ \theta_{k+j+1},\ldots,\theta_n)$.
4. Evaluate partial samples $(\theta_1^*,\ldots,\theta_k^*,\ \bar{\theta}_{k+1},\ldots,\bar{\theta}_{k+j},\ \theta_{k+j+1},\ldots,\theta_n)$ by solving the subproblems using math programming. Calculate the promising indexes. If the next most promising region is within the current most promising region, further partitioning; otherwise, backtracking. Go to step 1.

5.3.2.1 Sampling

The first step in applying the HNP–MP approach to IHLP problems is to determine a proper form of partial samples such that the capability of MIP solvers such as CPLEX can be fully leveraged to efficiently solve the small-scale subproblems associated with the partial samples. For the IHLP, we define partial samples as feasible solutions to the problem in the form of fixing a set of hubs to be closed (some z variables are fixed to 0), and no flow can move through these closed hubs. Figure 5.1 shows an example with seven hubs $r_1,\ldots,r_7,$ and four flows $f_1,\ldots,f_4;$ and in the partial solution, hubs r_1, r_4 and r_6 are closed.

Biased sampling techniques can be used to obtain partial samples with high quality. We develop a biased sampling procedure, called the linear programming (LP) solution-based sampling, for solving the IHLP problem. The procedure is described in Algorithm 5.3.

Algorithm 5.3 LP Biased Sampling

1. Obtain the LP solution. Denote the solution of variable z by z^*.
2. Calculate the sampling weights of variable z, based on the value of z^*. For $\forall k \in R$, define $W_k' = z_k^* + \varepsilon$ (ε is a very small nonnegative number). The sampling weights are then obtained by normalization

$$W(\text{hub } k \text{ is open}) = W_k = W_k' \Big/ \sum_{k' \in R} W_{k'}', \quad \forall k \in R.$$

3. Based on the sampling weights, generate partial samples.

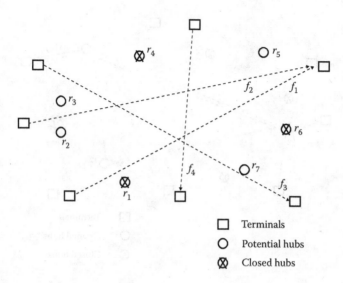

FIGURE 5.1
IHLP: a partial sample.

5.3.2.2 *Calculating the Promising Index*

To calculate the promising index, we first need to evaluate the partial samples generated in the sampling step. Given the partial sample, the remaining problem is a smaller subproblem that can be solved quickly using math programming. In Figure 5.1, hubs r_1, r_4 and r_6 are closed. By solving the corresponding subproblem, a complete sample solution can be obtained as shown in Figure 5.2. The elite samples obtained will be used to calculate the promising index and guide the partitioning/backtracking. For the IHLP, if only a fraction of the z variables are fixed (to 0) in a partial sample, the partial sample corresponds to a relatively small problem with the same structure as the original problem; if the partial samples are in the form of fixing all the z variables, the subproblem associated with each partial sample becomes a linear program. For both cases, standard math programming approaches can be used to evaluate partial samples efficiently.

5.3.2.3 *Partitioning, Backtracking, and Stopping*

If the most promising region needs to be further partitioned, we keep the current best sample in the next most promising region, which provides a set of partitioning variables. For the IHLP, each available partitioning variable can be used to partition current most promising region into two subregions (one subregion with the given hub open and the other with the given hub closed).

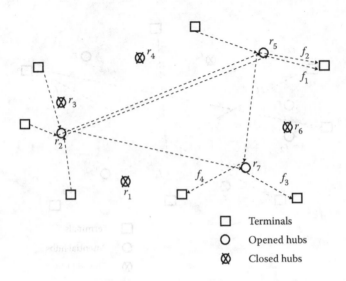

FIGURE 5.2
IHLP: a complete sample solution.

If backtracking is performed, then some constraints on the most promising region are relaxed. For the IHLP, by dropping some cuts that let certain hub(s) open, the next most promising region will include the current most promising region and the best sample obtained so far. The algorithm usually stops when the computational resource (e.g., time) reaches a predefined value. Other problem-depend stopping criteria can also be designed.

5.3.3 Computational Results

In Pi et al. (2008), 21 randomly generated cases were tested. The solutions generated by the HNP–MP approach were compared to those computed by CPLEX MIP solver and a Lagrangian relaxation (LR) approach in Geoffrion (1974), respectively. For these instances, the sizes of terminal locations range from 30 to 60, the sizes of candidate hubs range from 30 to 50, and the size of flows range from 150 to 300. Owing to the limited space, we refer to Pi et al. (2008) for detailed settings of each test case. The test results are shown in Table 5.1. In the table, LB is the best lower bound obtained using CPLEX MIP solver; CPLEX is the solution obtained using the CPLEX MIP solver; Gap CPLEX (%) is the optimality gap of CPLEX solution; LR is the solution obtained using the Lagrangian relaxation approach; Gap LR (%) is the optimality gap of LR solution compared to LB; HNP–MP is the solution obtained using the HNP–MP approach; Gap HNP-MP (%) is the optimality gap of HNP–MP solution compared to LB.

From Table 5.1, the solutions found by HNP–MP is superior to those found by LR and standard CPLEX MIP solver. Also, the HNP–MP solutions are

TABLE 5.1

IHLP Test Results

Case	LB	CPLEX	Gap CPLEX (%)	LR	Gap LR (%)	HNP–MP	Gap HNP–MP (%)
1	487,561	604,171	23.9	518,017	6.2	494,705	1.5
2	448,997	461,617	2.8	505,079	12.5	454,880	1.3
3	320,345	403,153	25.8	337,503	5.4	324,313	1.2
4	313,337	385,607	23.1	329,818	5.3	317,509	1.3
5	238,229	302,019	26.8	256,319	7.6	241,663	1.4
6	232,600	253,196	8.9	234,998	1	234,898	1
7	312,283	350,581	12.3	329,982	5.7	315,783	1.1
8	553,196	604,171	9.2	567,793	2.6	565,163	2.2
9	523,013	570,222	9	555,238	6.2	528,691	1.1
10	365,622	401,353	9.8	371,864	1.7	372,036	1.8
11	354,859	385,607	8.7	361,218	1.8	359,298	1.3
12	275,713	302,020	9.5	281,922	2.3	275,793	0
13	267,871	300,074	12	274,028	2.3	269,229	0.5
14	353,558	395,128	11.8	367,467	3.9	358,710	1.5
15	477,633	604,171	26.5	485,076	1.6	481,618	0.8
16	447,052	455,698	1.9	484,173	8.3	454,129	1.6
17	305,522	401,353	31.4	314,423	2.9	312,196	2.2
18	295,429	385,607	30.5	312,687	5.8	301,513	2.1
19	229,240	302,020	31.7	237,010	3.4	233,042	1.7
20	222,205	300,074	35	248,134	11.7	226,639	2
21	299,684	395,128	31.8	312,503	4.3	305,789	2
Average			18.2		4.8		1.4

very close to the optimality, owing to the facts that good sample solutions are quickly identified to speed up the search.

5.4 Nested Partitions for Production Planning

In this section, we review an implementation of NP in solving the multi-level capacitated lot-sizing problem with backlogging (ML-CLSB) published in Wu et al. (2011). ML-CLSB is a class of problems in production planning.

The ML-CLSB problem intends to plan production at each stage of a complex bill-of-materials (BOM) over a finite horizon. The objective is to minimize the total costs, including setup costs, inventory holding costs, and backlogging costs, while meeting all customer demands by the end of the planning horizon. Effective solutions of this problem is one important determinant of cost performance in Material Requirements Planning (MRP), Manufacturing Resource Planning (MRPII), and Enterprise Resource Planning (ERP) systems.

The ML-CLSB problems are notoriously difficult to solve due to their complexity. Past research has focused on simpler variants (single-level or uncapacitated or both) (Pochet and Wolsey, 1988; Ganas and Papachristos, 2005; Song and Chan, 2005; Mathieu, 2006). Heuristics and metaheuristics have also been popular in solving lot-sizing problems due to the difficulties encountered using exact algorithms (Kuik and Salomon, 1990; Ozdamar and Bozyel, 2000; Tang, 2004; Karimi et al., 2006; Akartunalı and Miller, 2009).

5.4.1 Mathematical Model

Multiple mathematical formulations exist in literature. They result in different performance when using different solution approaches. A formulation that generates tight lower bound and is suitable for using the NP method is presented in the following context:

Sets

- $T = \{1,\ldots,|T|\}$: set of time periods in the planning horizon
- $M = \{1,\ldots,|M|\}$: set of machines
- $I = \{1,\ldots,|I|\}$: set of items
- $endp \subset I$: set of end items
- $endp_i \subseteq endp, i \in I$: set of end items that utilize item i
- $\eta_i \subset I$: set of immediate successors of item i

Parameters

- $sc_i, i \in I$: setup cost for producing item i
- $ec_i, i \in I$: echelon inventory holding cost for one unit of item i per time period, which is the unit holding cost for item i minus the total holding cost of its immediate predecessors needed (Akartunalı and Miller, 2009)
- $bc_i, i \in I$: backlogging cost for one unit of item i per time period
- $ed_{it}, i \in I, t \in T$: echelon demand for item i in period t, which is the gross demand needed for item i in period t plus the amount of item i needed to produce its immediate successor (Akartunalı and Miller, 2009)
- $r_{ij}, i \in I, j \in I$: amount of item i needed to produce one unit of item j, where item j is one of the successors of item i
- $a_{im}, i \in I, m \in M$: production time required to produce one unit of item i on machine m
- $st_{im}, i \in I, m \in M$: setup time required for producing item i on machine m
- $C_{mt}, m \in M, t \in T$: available capacity of machine m in period t

Decision variables

- x_{itp}, $i \in I$, $t \in T$, $p \in T$: amount of item i produced in period t to satisfy demand in period p
- y_{it}, $i \in I$, $t \in T$: binary setup decision variables. $y_{it} = 1$ if production is setup for item i in period t; $y_{it} = 0$ otherwise

Formulation

min

$$\sum_{i \in I} \sum_{t \in T} sc_i \cdot y_{it} \qquad (5.13)$$

$$+ \sum_{i \in I} \sum_{t \in T} \sum_{p=t}^{T} ec_i \cdot (p-t) \cdot x_{itp} \qquad (5.14)$$

$$+ \sum_{i \in endp} \sum_{t \in T} \sum_{p=1}^{t-1} bc_i \cdot (t-p) \cdot x_{itp} \qquad (5.15)$$

Subject to

$$\sum_{p \in T} x_{ipt} = ed_{it}, \quad \forall i \in I, \ t \in T, \qquad (5.16)$$

$$\sum_{q=1}^{t} \sum_{p=t+1}^{T} x_{iqp} \geq \sum_{j \in n_i} r_{ij} \cdot \sum_{q=1}^{t} \sum_{p=t+1}^{T} x_{jqp}, \quad \forall i \in I \setminus endp, \ t \in T, \qquad (5.17)$$

$$\sum_{j \in endp_i} r_{ij} \cdot \sum_{q=t+1}^{T} \sum_{p=1}^{T} x_{jqp} = \sum_{q=t+1}^{T} \sum_{p=1}^{t} x_{iqp}, \quad \forall i \in I \setminus endp, \ t \in T, \qquad (5.18)$$

$$x_{itp} \leq ed_{ip} \cdot y_{it}, \quad \forall i \in I, \ t \in T, p \in T, \qquad (5.19)$$

$$\sum_{i \in T} \sum_{p \in T} a_{im} \cdot x_{itp} + \sum_{i \in I} st_{im} \cdot y_{it} \leq C_{mt}, \quad \forall m \in M, \ t \in T, \qquad (5.20)$$

x_{itp} are nonnegative, and y_{it} are binary.

Here, the objective function consists of the setup costs (Equation 5.13), the inventory holding costs (Equation 5.14), and the backlogging costs

(Equation 5.15). The constraints in Equation 5.16 ensure that the demands in each period are satisfied. The constraints in Equation 5.17 mean that echelon inventories for non-end items are no smaller than those of their corresponding successors. The constraints in Equation 5.18 guarantee that echelon backlog levels for non-end items are consistent with the backlog levels of their corresponding end items. The constraints in Equation 5.19 are setup forcing constraints. The constraints in Equation 5.20 enforce production capacity limits.

5.4.2 Lower and Upper Bound Guided Nested Partitions

In this section, a hybrid NP approach is introduced, called the lower-and-upper bound guided nested partitions (LugNP). The LugNP method focuses on solving mixed-integer programming models in the following form.

$$
\begin{aligned}
\min \quad & c_1 x + c_2 y \\
\text{s.t.} \quad & A_1 x + A_2 y \geq b \quad (P) \\
& x \in R^n, y \in \{0,1\}^m
\end{aligned}
$$

Problem (P) represents typical mixed binary integer programming models, which our ML-CLSB model falls into. The difficulties in solving these problems arise as a result of the existence of binary variables y. The basic idea of the LugNP method is to efficiently fix a subset of the binary variables that leads to a restricted version of problem (P) that is easier to solve. It starts with quickly finding a feasible solution using certain heuristics. This idea is inherited from the HNP–MP approach presented in Section 5.3.

Denote four subsets of the binary variable y as qf, qp, qs, and qm, representing the index sets of binary variables that are fixed in the current iteration, currently used for partitioning, currently used for sampling, and the free variables, respectively. Given qf, qp, and qs, the restricted version of problem (P) is denoted as (P^S), where y_p are known for all $p \in qf \cup qp \cup qs$. Solving problem (P^S) is to obtain unknown variables x and y_p, $p \in qm$.

LugNP relies on a lower bound and an upper bound updated in every iteration. To obtain such a lower bound, a lower bounding technique, such as linear programming relaxation, is desired and denoted as L. Meanwhile, a heuristic or math programming, denoted as H, is used to obtain the upper bounds. With the aforementioned notations, the LugNP method is described in Algorithm 5.4.

Algorithm 5.4 LugNP Algorithm

1. **Initialization.** Solve (P) using L and H to obtain a lower bound LB and an initial feasible solution UB. The corresponding binary variables in LB and UB are denoted as $\underline{y}$ and $\bar{y}$. Set $k = 1$, $qf = \varnothing$, $qp = \varnothing$, $qs = \varnothing$, and the current most promising region $\sigma(k) = \Theta$.

2. **Select partitioning and sampling variables.** For each free binary variable in qm, compute the following probability based on $\underline{y}$ and $\bar{y}$.

$$Pr_p = \frac{\rho^{1-|\bar{y}_p - \underline{y}_p|}}{\rho} \Bigg/ \sum_{p' \in qm} \frac{\rho^{1-|\bar{y}_{p'} - \underline{y}_{p'}|}}{\rho}, \quad \forall p \in qm.$$

Here $\rho \geq 1$ so that a larger value of ρ leads to a higher probability of selecting a variable with a smaller gap between its corresponding $\bar{y}$ and $\underline{y}$. Sort all binary variables in qm in the descending order of Pr_p. Select the top one or several binary variables in qm as the next partitioning variables (qp), and the next one or several binary variables as the sampling variables (qs).

3. **Partitioning.** Partition the current most promising region into M subregions and one surrounding region, based on the selected partitioning variables qp.

4. **Partial sampling.** Within each subregion $\sigma_j(k)$ (binary variables in qf and qp fixed), randomly sample the variables in qs and construct the corresponding Nj subproblems

$$P^S_{kj1}, \ldots, P^S_{kjN_j}.$$

5. **Estimate promising index.** Within each subregion $\sigma_j(k)$, solve all subproblems using the lower bound technique L to obtain lower bounds

$$L_0\left(P^S_{kj1}\right), \ldots, L_0\left(P^S_{kjN_j}\right).$$

Determine the subproblem with the best lower bound, whose index is denoted by $\hat{i}_{kj}$ as

$$\hat{i}_{kj} = \arg\min_{i \in \{1, \ldots, N_j\}} L_0\left(P^S_{kji}\right).$$

The promising index of $\sigma_j(k)$ is then determined by solving the subproblem $\left(P^S_{kji}\right)$, using the heuristic H, as

$$H_0\left(P^S_{kj(\hat{i}_{kj})}\right).$$

6. **Determine the next most promising region and move.** The next most promising region, whose index is denoted by $\hat{j}_k$, is chosen as

$$\hat{j}_k = \arg\min_{j \in \{1, \ldots, M+1\}} H_0\left(P^S_{kj(\hat{i}_{kj})}\right).$$

Update $\bar{y}$ using the solution of $H_0\left(P^S_{k(\hat{j}_k)(\hat{i}_{kj})}\right)$ and $\underline{y}$ using the solution of $L_0\left(P^S_{k(\hat{j}_k)(\hat{i}_{kj})}\right)$.

Stop if all binary variables are in qf or time limit is reached. Otherwise, either further partitioning or backtracking based on the next most promising region. Go to step 1.

There are two key elements in the LugNP algorithm.

1. In each iteration, the new partitioning and sampling variables are chosen based on the current best lower bound and upper bound. For a given binary variable in y, the smaller the gap between the variable's values in $\bar{y}$ and $\underline{y}$, the higher probability it is chosen as a partitioning and sampling variable. Partitioning variables have higher priorities than sampling variables.

2. When estimating the performance of each subregion, we do not need to solve all partial sample, that is, subproblems $P^S_{kj1}, \ldots, P^S_{kjN_j}$, using the heuristic algorithm H. Instead, all the subproblems are solved using the lower bound technique L, which is generally much faster than H. For example, H can be a MIP solver, while L can be a LP solver. Only the subproblem with the best lower bound is solved using H to obtain a feasible solution in this subregion. Hence, the computational time is greatly shortened.

To solve the ML-CLSB problem, we choose LP relaxation to be the lower bound technique L and a relax-and-fix heuristic to be the upper bound heuristic H. The basic idea of the relax-and-fix algorithm is to keep only a subset of the binary variables at a time, and relax the binary requirement on others. The resulting problem, which is now smaller, is then solved using a MIP solver. Once we fix this subset of the binary variables, the similar procedure is repeated until the entire subproblem is solved (Wu et al., 2011).

5.4.3 Computational Results

In Wu et al. (2011), 12 sets of instances were tested. The first four test sets, denoted by $\bar{A}+, \bar{B}+, \bar{C}, \bar{D}$, were modified based on the benchmarks published in Tempelmeier and Derstroff (1996) and Stadtler (2003) to permit backlogging. They contain 120, 312, 144, and 79 test cases, respectively. The second four test sets, denoted by $\bar{\bar{A}}+, \bar{\bar{B}}+, \bar{\bar{C}}, \bar{\bar{D}}$, were modified based on the first four test sets to allow a more significant role of backlogging. The last four test sets, denoted by SET1, SET2, SET3, and SET4, were adopted from Akartunalı and Miller (2009). They contain 30 instances each. We refer to Wu et al. (2011) for detailed description of test instances and settings.

The LugNP method is compared with the CPLEX MIP solver (branch-and-cut) and a heuristic proposed in Akartunalı and Miller (2009). The average

TABLE 5.2

ML-CLSB Test Results

Data	Heuristic (%)	CPLEX (%)	LugNP (%)	Imp_Heuristic (%)	Imp_CPLEX (%)
$\bar{A}+$	30.21	32.24	24.76	18.03	23.20
$\bar{B}+$	29.42	34.26	23.86	18.90	30.36
$\bar{C}$	47.84	39.75	29.10	39.16	26.79
$\bar{D}$	28.51	29.72	14.89	47.78	49.91
$\bar{\bar{A}}+$	28.98	30.82	24.92	13.99	19.12
$\bar{\bar{B}}+$	34.30	34.69	28.57	16.69	17.63
$\bar{\bar{C}}$	46.79	40.28	29.92	36.05	25.73
$\bar{\bar{D}}$	43.31	79.95	32.73	24.43	59.06
SET1	14.36	19.59	14.28	0.54	27.07
SET2	8.84	12.43	8.82	0.22	28.98
SET3	206.04	248.13	172.69	16.18	30.40
SET4	108.68	123.60	105.59	2.84	14.57

performance on each set of test instances are shown in Table 5.2. The optimality gaps of solutions generated by the heuristic, CPLEX, and LugNP are shown, where the lower bound is obtained by the LP relaxation. "Imp_Heuristic (%)" and "Imp_CPLEX (%)" represent percentage of improvements of LugNP solutions compared to the heuristic and CPLEX solutions, respectively. The table shows that LugNP outperforms both the heuristic and the CPLEX MIP solver. On average, LugNP improves the solution quality by 19.84% and 27.45% from heuristic and CPLEX, respectively.

5.5 Conclusions

With the increasingly intensified competition, companies have paid much attention to build effective and robust supply chains. Supply chain management is the collection of approaches to achieve that goal. In recent years, optimization has gained more and more recognition in supply chain management because of its rigor and power to pursue cost reduction and efficiency improvement. As one can imagine, many of these problems emerging in supply chain optimization are large scale and notoriously difficult to solve. The nested partitions method, along with other exact and heuristic algorithms, is developed to tackle these large-scale optimization problems.

In this chapter, the nested partitions method is reviewed and its global convergence is presented. It is a partitioning and sampling based global random search algorithm. The real power of nested partitions lies in its flexibility to incorporate other heuristics or exact algorithms. The NP framework guides the optimization search to be focused on the subsets of the solution space where the optimum is most likely to exist. Within each subset, the incorporated heuristic or exact algorithm quickly explores and searches for the local optimum. To demonstrate the hybrid NP approaches, we review two important supply chain problems solved by NP. One example is the intermodal hub location problem, a special case of facility location problems. The mathematical model is provided, and a hybrid NP and math programming (HNP–MP) approach is introduced to solve the problem. Test results show that the HNP–MP approach is superior to the CPLEX MIP solver and a Lagrangian relaxation method. Another example is the multilevel capacitated lot-sizing problem with backlogging (ML-CLSB), which is a complex production planning problem. A binary integer programming (BIP) model is provided, and a lower and upper bound guided NP (LugNP) method is proposed to deal with such a BIP model. The LugNP method utilizes lower bounds and upper bounds to determine a good set of partitioning and sampling variables, as well as the next most promising region. Computational tests on 12 sets of instances show 19.84% of improved performance over a heuristic proposed to solve ML-CLSB problems and 27.45% of improvement over the standard CPLEX MIP solver.

In short, this chapter reviews the nested partitions method and its applications to supply chain management. More importantly, we would like to place more emphasis on supply chain optimization and hope to inspire more advancements in such an important area.

References

Aboolian, R., T. Cui, and Z.-J. M. Shen. An efficient approach for solving reliable facility location models. *INFORMS Journal on Computing*, 25(4): 720–729, 2013.

Akartunalı, K. and J. Miller. A heuristic approach for big bucket multi-level production planning problems. *European Journal of Operational Research*, 193(2): 396–411, 2009.

Amiri, A. Designing a distribution network in a supply chain system: Formulation and efficient solution procedure. *European Journal of Operational Research*, 171(2): 567–576, 2006.

Arreola-Risa, A. and G. A. DeCroix. Inventory management under random supply disruptions and partial backorders. *Naval Research Logistics*, 45(7): 687–703, 1998.

Bertsimas, D. and C.-P. Teo. From valid inequalities to heuristics: A unified view of primal-dual approximation algorithms in covering problems. *Operations Research*, 46(4): 503–514, 1998.

Bertsimas, D. and A. Thiele. A robust optimization approach to supply chain management. In D. Bienstock and G. Nemhauser (eds.), *Integer Programming and Combinatorial Optimization* (pp. 86–100). New York: Springer Science+Business Media, 2004.

Bräsy, O. and M. Gendreau. Vehicle routing problem with time windows, Part I: Route construction and local search algorithms. *Transportation Science*, 39(1): 104–118, 2005a.

Bräsy, O. and M. Gendreau. Vehicle routing problem with time windows, Part II: Metaheuristics. *Transportation Science*, 39(1): 119–139, 2005b.

Campbell, J. F. Integer programming formulations of discrete hub location problems. *European Journal of Operation Research*, 72: 387–405, 1994.

Campbell, A. M. and M. Savelsbergh. Efficient insertion heuristics for vehicle routing and scheduling problems. *Transportation Science*, 38(3): 369–378, 2004.

Chen, X., M. Sim, D. Simchi-Levi, and P. Sun. Risk aversion in inventory management. *Operations Research*, 55(5): 828–842, 2007.

Chen, W., L. Pi, and L. Shi. Nested partitions and its applications to the intermodal hub location problem. In W. Chaovalitwongse, K. Furman, and P. Pardalos (eds.), *Optimization and Logistics Challenges in the Enterprise* (pp. 229–251). New York: Springer Science+Business Media, 2009.

Chen, W., J. Song, and L. Shi. Optimizing local pickup and delivery with uncertain loads. In *Proceedings of the 2011 Winter Simulation Conference* (pp. 4246–4256). Phoenix, AZ. December 11–14, 2011.

Chen, W., J. Song, and L. Shi. Data mining-based dispatching system for solving the local pickup and delivery problem. *Annals of Operations Research*, 203(1): 351–370, 2013.

Chopra, S. Designing the distribution network in a supply chain. *Transportation Research Part E*, 39(2): 123–140, 2003.

Cormen, T. H., C. E. Leiserson, R. L. Rivest, and C. Stein. *Introduction to Algorithms*. Cambridge, MA: MIT Press and New York: McGraw-Hill, 1990.

Cui, T., Y. Ouyang, and Z.-J. M. Shen. Reliable facility location design under the risk of disruptions. *Operations Research*, 58(4): 998–1011, 2010.

Cvijovíc, D. and J. Klinowski. Taboo search: An approach to the multiple minima problem. *Science*, 267(5198): 664–666, 1995.

Daniel, V. R. G., Jr. and L. N. V. Wassenhove. The reverse supply chain. *Harvard Business Review*, 80(2): 25–26, 2002.

Daskin, M. S., L. V. Snyder, and R. T. Berger. Facility location in supply chain design. In A. Langevin and D. Riopel (eds.), *Logistics Systems: Design and Optimization* (pp. 39–65). New York: Springer Science+Business Media, 2005.

De Boer, P. T., D. P. Kroese, S. Mannor, and R. Y. Rubinstein. A tutorial on the cross-entropy method. *Annals of Operations Research*, 134: 19–67, 2005.

Dorigo, M. *Optimization, Learning and Natural Algorithms*. PhD thesis, Politecnico di Milano, Italy, 1992.

Dorigo, M., G. D. Caro, and L. M. Gambardella. Ant algorithms for discrete optimization. *Artificial Life*, 5(2): 137–172, 1999.

Drexl, A. and A. Kimms. Lot sizing and scheduling—Survey and extensions. *European Journal of Operational Research*, 99(2): 221–235, 1997.

Fisher, M. L. The Lagrangian relaxation method for solving integer programming problems. *Management Science*, 50(12): 1861–1871, 2004.

Ganas, I. and S. Papachristos. The single-product lot-sizing problem with constant parameters and backlogging: Exact results, a new solution, and all parameter stability regions. *Operations Research*, 53(1): 170–176, 2005.

Geoffrion, A. Lagrangian relaxation for integer programming. *Mathematical Programming Study 2*, 2: 82–114, 1974.

Glover, F. W. and M. Laguna. *Tabu Search*. Boston: Kluwer Academic, 1997.

Goldberg, D. E. *Genetic Algorithm in Search, Optimization and Machine Learning*. Reading, MA: Addison-Wesley, 1989.

Gupta, A. and C. D. Maranas. Managing demand uncertainty in supply chain planning. *Computers & Chemical Engineering*, 27(8–9): 1219–1227, 2003.

Hall, N. G. and D. S. Hochbaum. A fast approximation algorithm for the multi covering problem. *Discrete Applied Mathematics*, 15(1): 35–40, 1986.

Hall, N. G. and C. N. Potts. Supply chain scheduling: Batching and delivery. *Operations Research*, 51(4): 566–584, 2003.

Hu, J., M. C. Fu, and S. I. Marcus. A model reference adaptive search method for global optimization. *Operations Research*, 55: 549–568, 2007.

Karimi, B., S. M. T. F. Ghomi, and J. M. Wilson. A tabu search heuristic for solving the clsp with backlogging and set-up carry-over. *Journal of the Operational Research Society*, 57(2): 140–147, 2006.

Kennedy, J. and R. Eberhart. Particle swarm optimization. In *Proceedings of 1995 IEEE International Conference on Neural Networks*, Vol. 4 (pp. 1942–1948), Perth, WA. November 27–December 1, 1995.

Kirkpatrick, S., C. D. Gelatt, and M. P. Vecchi. Optimization by simulated annealing. *Science*, 220(4598): 671–680, 1983.

Klose, A. and A. Drexl. Facility location models for distribution system design. *European Journal of Operational Research*, 162(1): 4–29, 2005.

Kreipl, S. and M. Pinedo. Planning and scheduling in supply chains: An overview of issues in practice. *Production and Operations Management*, 13(1): 77–92, 2004.

Kuik, R. and M. Salomon. Multi-level lot-sizing problem: Evaluation of a simulated annealing heuristic. *European Journal of Operational Research*, 45(1): 25–37, 1990.

Lemarechal, C. Lagrangian relaxation. In M. Jünger and D. Naddef (eds.), *Computation Combinatorial Optimization* (pp. 112–156). *Lecture Notes in Computer Science*, Vol. 2241. Berlin and Heidelberg: Springer-Verlag, 2001.

Maravelias, C. T. and C. Sung. Integration of production planning and scheduling: Overview, challenges and opportunities. *Computers & Chemical Engineering*, 33(12): 1919–1930, 2009.

Mathieu, V. V. Linear programming extended formulations for the single item lot sizing problem with backlogging and constant capacity. *Mathematical Programming*, 108(1): 53–77, 2006.

Miller, T. *Hierarchical Operations and Supply Chain Planning*. New York: Springer Science+Business Media, 2002.

Mitchell, M. *An Introduction to Genetic Algorithms*. Cambridge, MA: MIT Press, 1998.

Nemhauser, G. L. and L. A. Wolsey. *Integer and Combinatorial Optimization*. New York: Wiley-Interscience, 1999.

O'Kelly, M. E. and D. L. Bryan. Hub location with flow economies of scale. *Transportation Research Part B*, 32(8): 605–616, 1998.

Ólafsson, S. and J. Yang. Intelligent partitioning for feature selection. *INFORMS Journal on Computing*, 17(3): 339–355, 2005.

Ozdamar, L. and M. A. Bozyel. The capacitated lot sizing problem with overtime decisions and setup times. *IIE Transactions*, 32(11): 1043–1057, 2000.

Pi, L., Y. Pan, and L. Shi. Hybrid nested partitions and mathematical programming approach and its applications. *IEEE Transactions on Automation Science and Engineering*, 5(4): 573–586, 2008.

Pinedo, M. *Scheduling: Theory, Algorithms, and Systems*, 3rd ed. New York: Springer Science+Business Media, 2008.

Pinedo, M. and X. Chao. *Operations Scheduling with Applications in Manufacturing and Services*. New York: McGraw-Hill/Irwin, 1999.

Pishvaee, M. S., M. Rabbani, and S. A. Torabi. A robust optimization approach to closed-loop supply chain network design under uncertainty. *Applied Mathematical Modelling*, 35(2): 637–649, 2011.

Pochet, Y. and L. A. Wolsey. Lot-size models with backlogging: Strong reformulations and cutting planes. *Mathematical Programming*, 40(3): 317–335, 1988.

Poli, R., J. Kennedy, and T. Blackwell. Particle swarm optimization. *Swarm Intelligence*, 1(1): 33–57, 2007.

Powell, W. B., J. Shapiro, and H. P. Simao. An adaptive, dynamic programming algorithm for the heterogeneous resource allocation problem. *Transportation Science*, 36(2): 231–249, 2002.

Puterman, M. L. *Markov Decision Processes: Discrete Stochastic Dynamic Programming*. New York: Wiley-Interscience, 2005.

Racunica, I. and L. Wynter. Optimal location of intermodal freight hubs. *Transportation Research Part B*, 39: 453–477, 2005.

Roy, B. V., D. P. Bertsekas, Y. Lee, and J. N. Tsitsiklis. A neuro-dynamic programming approach to retailer inventory management. In *Proceedings of the 36th IEEE Conference on Decision and Control*, Vol. 4 (pp. 4052–4057). San Diego, CA. December 10–12, 1997.

Rubinstein, R. Y. and D. P. Kroese. *The Cross-Entropy Method: A Unified Approach to Combinatorial Optimization, Monte-Carlo Simulation, and Machine Learning*. New York: Springer Science+Business Media, 2004.

Sandholm, T., D. Levine, M. Concordia, P. Martyn, R. Hughes, J. Jacobs, and D. Begg. Changing the game in strategic sourcing at Procter & Gamble: Expressive competition enabled by optimization. *Interfaces*, 36(1): 55–68, 2006.

Santoso, T., S. Ahmed, M. Goetschalckx, and A. Shapiro. A stochastic programming approach for supply chain network design under uncertainty. *European Journal of Operational Research*, 167(1): 96–115, 2005.

Shi, L. and S. Ólafsson. Nested partitions method for global optimization. *Operations Research*, 48(3): 390–407, 2000a.

Shi, L. and S. Ólafsson. Nested partitions method for stochastic optimization. *Methodology and Computing in Applied Probability*, 2(3): 271–291, 2000b.

Shi, L. and S. Men. Optimal buffer allocation in production lines. *IIE Transactions*, 35: 1–10, 2003.

Shi, L., S. Ólafsson, and Q. Chen. An optimization framework for product design. *Management Science*, 47(12): 1681–1692, 2001.

Simchi-Levi, D., P. Kaminsky, and E. Simchi-Levi. *Designing and Managing the Supply Chain: Concepts, Strategies and Case Studies*, 3rd ed. New York: McGrawHill/Irwin, 2008.

Song, Y. Y. and G. H. Chan. Single item lot-sizing problems with backlogging on a single machine at a finite production rate. *European Journal of Operational Research*, 161(1): 191–202, 2005.

Stadtler, H. Multilevel lot sizing with setup times and multiple constrained resource: Internally rolling schedules with lot-sizing windows. *Operations Research*, 51(3): 487–502, 2003.

Tang, O. Simulated annealing in lot sizing problems. *International Journal of Production Economics*, 88(2): 173–181, 2004.

Tempelmeier, H. and M. Derstroff. A Lagrangian-based heuristic for dynamic multi-level multi-item constrained lotsizing with setup times. *Management Science*, 42(5): 738–757, 1996.

van Laarhoven, P. J. M. and E. H. L. Aarts. *Simulated Annealing: Theory and Applications*. Norwell, MA: Kluwer Academic, 1987.

Voß, S. and D. L. Woodruff. *Introduction to Computational Optimization Models for Production Planning in a Supply Chain*, 2nd ed. New York: Springer Science+ Business Media, 2006.

Wolsey, L. A. *Integer Programming*. New York: Wiley-Interscience, 1998.

Wu, T., L. Shi, J. Geunes, and K. Akartunalı. An optimization framework for solving capacitated multi-level lot-sizing problems with backlogging. *European Journal of Operational Research*, 214(2): 428–441, 2011.

Xu, H., Z.-L. Chen, S. Rajagopal, and S. Arunapuram. Solving a practical pickup and delivery problem. *Transportation Science*, 37(3): 347–364, 2001.

Zhang, H. H., L. Shi, R. Meyer, D. Nazareth, and W. D'Souza. Solving beam angle selection and dose optimization simultaneously via high-throughput computing. *INFORMS Journal on Computing*, 21(3): 427–444, 2009.

Section III

Inventory Management in the Supply Chain

6

A Schedule-Based Formulation for the Cyclic Inventory Routing Problem

Zhe Liang, Rujing Liu, and Wanpracha Art Chaovalitwongse

CONTENTS

ABSTRACT   In this chapter, we study a cyclic inventory routing problem (CIRP). The traditional exact methods for the inventory routing problem (IRP) use an arc-based formulation (also known as two-index flow formulation), in which a variable represents a possible vehicle flow between a pair of customers. In this research, we propose a schedule-based model (SBM), in which a variable represents a possible one-day schedule for any vehicle. This model can be considered a Dantzig–Wolfe decomposition of the arc-based model. We also propose a set of new valid inequalities to tighten the linear relaxation bound of SBM. To solve SBM efficiently, we develop a column generation algorithm in which only attractive vehicle schedules are generated. Our computational results on five real-life test cases show that the linear programming (LP) relaxation of SBM is tight. SBM can obtain near-optimal solutions to very large real-life test cases within a reasonable time, and average integer programming (IP)–LP gaps of SBM are within 5% for maximum-level (ML) policy and 7% for order-up-to-level (OU) policy, respectively.

KEY WORDS: *column generation, cyclic schedule, integer programming, inventory routing problem, valid inequality.*

6.1 Introduction

The inventory routing problem (IRP) integrates three major decisions— inventory management, vehicle routing, and delivery scheduling—in the supply chain management. In the classical IRP, a fleet of homogeneous capacitated vehicles located at a central depot is used to serve a group of customers over multiple periods. In each period, the vehicles start and end the routes at the depot while the total inventory carried by each vehicle cannot exceed the vehicle capacity. The objective of IRP is to construct a replenishment schedule for each customer and a set of vehicle routes while minimizing the total inventory and transportation cost. There are two commonly used inventory replenishment policies: maximum-level (ML) policy and order-up-to-level (OU) policy. In the ML policy, the inventory delivered to a customer can be any positive value as long as the inventory at the customer does not exceed its maximum inventory capacity. In contrast, in the OU policy, whenever a customer is visited, the inventory delivered has to fullfill the customer's inventory capacity. In the last three decades, numerous exact solution methods and heuristics have been proposed for IRP. For a complete review of the theoretical and industrial development of IRP, we refer interested readers to two recent survey papers by Andersson et al. (2010) and Coelho et al. (2014).

IRP is closely related to the well-known capacitated vehicle routing problem (VRP). The traditional model for VRP is an arc-based formulation (also known as the two-index flow formulation; Toth and Vigo, 2014). However, recent advances in VRP suggest a very promising modeling direction, in which VRP is formulated as a set partitioning problem and each variable represents a unique vehicle route. The route-based model can optimally solve some very hard test cases that previously could not be solved by the arc-based model (Fukasawa et al., 2006; Baldacci et al., 2011; Pecin et al., 2014).

On the other hand, the most promising exact methods for IPR are those of Archetti et al. (2007) for the ML policy and Solyli and Sural (2011) for the

OU policy, which are all arc-based models. Adulyasak et al. (2014) performed an extensively computational study on the aforementioned methods. It is shown that these methods can solve only test cases with up to 35 customers, 3 periods, and 3 vehicles optimally. If the number of periods increases from 3 to 6, the size of the optimally solvable test cases decreases to 15 customers with 2 vehicles. Because the formulations proposed in Archetti et al. (2007), Solyli and Sural (2011), and Adulyasak et al. (2014) are arc-based models, branch-and-cut has to be used for optimal integer programming (IP) solutions. However, one of the major challenges is that the linear programming (LP) relaxations of these formulations are not tight. For example, Adulyasak et al. (2014) reported that when inventory cost is low, most gaps between the root note LP solution and the best IP solution range from 10% to 20%, and this root node IP–LP gap increases with the problem size. Consequently, the poor root note LP leads to a very long computational time.

Enlightened by the advances in VRP, in this chapter, we propose a schedule-based model for the cyclic inventory routing problem (CIRP) for both ML and OU policies. The variable in a schedule-based model represents a possible 1-day schedule for any vehicle. Then we develop a set of valid inequalities to tighten the linear relaxation of the model. Because the number of possible schedules increases exponentially with the number of customers, we also develop a column generation algorithm to solve the LP relaxation efficiently. Our computational results over five real-life large test cases show that the model proposed is competitive with both the exact methods of the ML policy of Archetti et al. (2007) and Adulyasak et al. (2014) and the OU policy of Solyli and Sural (2011). The proposed method can obtain very good IP solutions to the very large test case (up to 67 customers, 7 periods, and 16 vehicles) within 5% optimality in 4 hours of computational time.

This chapter is organized as follows. In Section 6.2, we provide a literature review on the CIRP. In Section 6.3, we first present an arc-based model, which is similar to the one in Archetti et al. (2007). Then we propose a schedule-based model (SBM), which can be viewed as a Dantzig–Wolfe decomposition of the arc-based model. We present the solution method for SBM in detail in Section 6.4. We also extend the schedule-based model for some real-life considerations in Section 6.5. Computational results on real-life test cases are reported in Section 6.6. Finally, conclusions and future research directions are given in Section 6.7.

6.2 Literature Review

Recently with the evolution of computational capability, researchers have been able to develop better solution approaches for many different IRP variations, such as stochastic IRP (Kleywegt et al., 2002, 2004; Adelman, 2004;

Hvattum et al., 2009), robust IRP (Solyli et al., 2012), maritime IRP (Gronhaug et al., 2010; Engineer et al., 2012), and consistent IRP (Coelho et al., 2012), to name but a few. Among them, the CIRP is used to find an inventory replenishment and vehicle routing schedule that are repeatable over a perpetual time horizon. CIRP is especially important and practical for customers with deterministic and periodic demands, that is, the deterministic demand pattern repeats itself periodically (e.g., weekly or monthly). Because the customer's demands are periodical and deterministic, a cyclic schedule is highly desirable. The studies of Anily and Federgruen (1990) and Gallego and Simchi-Levi (1990) are among the first to consider CIRP. Anily and Federgruen (1990) analyzed a class of "fixed partition policies," in which customers are grouped into regions, all of the customers within each region are served together by one vehicle, and different regions are served independently and separately. Gallego and Simchi-Levi (1990) showed that the long-run effectiveness of a direct shipping strategy is at least 94% effective whenever the Economic Lot Size is at least 71% of vehicle capacity. In the last decades, CIRP has drawn more and more attention from both industry and academia. Aghezzaf et al. (2006) and Raa and Aghezzaf (2008, 2009) studied a CIRP with constant demand rate. The length of the complete replenishment cycle is one of the decisions in the resulting schedule. Aghezzaf et al. (2006) proposed a mixed-integer programming model and a column generation solution algorithm to solve this problem, and Raa and Aghezzaf (2008) proposed a heuristic that is capable of handling more real-life constraints. Raa and Aghezzaf (2009) extended their previous work by integrating several heuristics into the column generation framework. Aghezzaf et al. (2012) and Vansteenwegen and Mateo (2014) proposed a mixed-integer programming model and an iterative local search algorithm to a single-vehicle CIRP, which raised a pricing subproblem for the column generation proposed by Aghezzaf et al. (2006). Zhao et al. (2008) studied a three-echelon logistic system containing a supplier, a central warehouse, and a group of retailers with constant demand rate. They first partitioned all the retailers into multiple regions as in Anily and Federgruen (1990) and Gallego and Simchi-Levi (1990), and then applied a power-of-two replenishment strategy for all the retailers in a region. They also proposed a variable large neighborhood search heuristic to optimize the partition of the retailers. Chan et al. (2013) studied a partition-based periodic policy, in which the retailers are partitioned into regions, and vehicles can serve all the retails in one or multiple regions at the same time. It is shown that a partition-based policy has the worst-cast asymptotic performance of 1.202 with respect to the best possible policy. Ekici et al. (2015) proposed an interactive clustering-based constructive heuristic to solve the CIRP in two stages: clustering and delivery schedule generation. It is worth mentioning that the CIRP they considered is different from all of those in the previous study because the number of periods in the cyclic schedule is predefined.

The CIRP we studied is similar to the one in Ekici et al. (2015). We assumed the number of periods in the cyclic schedule is predefined, for example,

7 days a week. Furthermore, instead of assuming a constant demand rate at each customer, we assume the demand rate for each customer is deterministic and periodically repeated. Therefore, the demand rates at different periods of the cycle might be different for the same customer. This is appropriate especially when the demand rate has seasonality.

6.3 Problem Definition and Formulation

The CIRP can be described as follows. There is a single depot, a set of customers, and a set of vehicles. Each customer has a deterministic demand rate in each period, and a customer can be visited at most once per period. The number of vehicles available is predefined, and a vehicle departs from and arrives at the depot in each period and can visit multiple customers as long as the inventory carried for these customers is less than the vehicle capacity. CIRP aims to find a schedule, lasting a predefined number of periods, for replenishing customers' inventory such that the long-term transportation cost is minimized. For each customer, the inventory at the end of the last period is equal to the inventory at the beginning of the first period, so that the replenishment schedule for all customers is repeatable over a perpetual time horizon. No stock out is allowed for any customers in CIRP.

6.3.1 Arc-Based Model

To facilitate our discussion, we define the following notations.

Sets, parameters, and constants

M: the set of customers, indexed by i and j

o: the depot where all vehicles depart from and arrive at

T: the set of planning periods, indexed by t

d_{it}: the demand for customer $i \in M$ in period $i \in T$

c_{ij}: the travel distance/cost between location i and j, where $i, j \in M \cup \{o\}$

B_i: the maximum inventory that customer i can hold

K: the set of vehicles available, indexed by k

H: the vehicle capacity

Variables

x_{ijkt}: the binary variable indicating whether vehicle $k \in K$ travels from customer i to customer j in period t. $x_{ijkt} = 1$ if vehicle k travels from customer i to j in period $t \in T$, and 0 otherwise

y_{it}: the inventory for customer $i \in M$ at the end of period $t \in T$

u_{ikt}: the binary variable indicating whether vehicle k visits customer i in period t. If $u_{ikt} = 1$, the customer is visited and 0 otherwise

v_{ikt}: the nonnegative variable representing the delivery amounts to customer i by vehicle k in period $t \in T$

The basic model (BM) can be formulated as follows:

$$\min \sum_{t \in T} \sum_{k \in K} \sum_{i,j \in M} c_{ij} x_{ijkt} \tag{6.1}$$

$$s.t. \sum_{j \in M} x_{ijkt} - \sum_{j \in M} x_{jikt} = 0 \quad \forall i \in M, \forall k \in K, \forall t \in T, \tag{6.2}$$

$$\sum_{j \in M} x_{m_0 jkt} = 1 \quad \forall k \in K, \forall t \in T, \tag{6.3}$$

$$\sum_{j \in M} x_{jm_0 kt} = 1 \quad \forall k \in K, \forall t \in T, \tag{6.4}$$

$$u_{ikt} \geq x_{ijkt} \quad \forall i, j \in M, \forall k \in K, \forall t \in T, \tag{6.5}$$

$$\sum_{k \in K} u_{ikt} \leq 1 \quad \forall i \in M, \forall t \in T, \tag{6.6}$$

$$v_{ikt} \leq \min\{B_i, H\} u_{ikt} \quad \forall i \in M, \forall k \in K, \forall t \in T \tag{6.7}$$

$$\sum_{i \in M} v_{ikt} \leq H \quad \forall k \in K, \forall t \in T, \tag{6.8}$$

$$y_{it} = y_{it-1} + \sum_{k \in K} v_{ikt} - d_{it} \quad \forall i \in M, \forall t \in T, \tag{6.9}$$

$$y_{it} + \sum_{k \in K} v_{ikt} \leq B_i \quad \forall i \in M, \forall k \in K, \tag{6.10}$$

$$x_{ijkt}, u_{ikt} \in \{0,1\}, y_{it}, v_{ikt} \geq 0 \quad \forall i, j \in M, \forall k \in K, \forall t \in T. \tag{6.11}$$

The objective in Equation 6.1 minimizes the total travel cost of all vehicles during the complete planning period. The constrains in Equations 6.2 through 6.4 are the flow balance constraints for each vehicle. The constraints in Equation 6.5 imply that if arc ij is traveled by a vehicle, customer i must be visited. The constraints in Equation 6.6 ensure that a customer can be visited at most once per period. The constraints in Equation 6.7 ensure the delivery amount for any customer i is less than or equal to the minimum of the vehicle capacity and the customer's inventory capacity. The constraints in Equation 6.8 ensure that the total delivery amount for the customers from vehicle k must be less than or equal to the vehicle capacity. The constraints in Equation 6.9 represent the inventory balance constraints for each customer during the complete planning horizon. Here we assume $t - 1$ is equal to $|T|$ when $t = 0$, which means that the inventory at the end of the planning horizon is equal to the inventory at the beginning of the planning horizon, so that the resulting schedule is repeatable from one planning horizon to the next. The constrains in Equation 6.10 ensure that the customer's inventory capacity cannot be exceeded. The constraints in Equation 6.11 are the binary and nonnegative constraints for variables.

6.3.2 Schedule-Based Model

The solutions provided by BM might contain sub-tours, and we need to add sub-tour elimination constraints to BM iteratively to find a feasible solution. This procedure could be very time consuming. Therefore, it is natural to decompose the BM using Dantzig–Wolfe decomposition, such that each variable in the model representing a feasible vehicle schedule in a single period. In the schedule-based model (SBM), a feasible vehicle schedule is a one-day trip that departs from and arrives at the depot and visits a sequence of customers. The total inventory delivered to the customers must obey the vehicle capacity constraint. As we can see, each vehicle schedule contains two types of information: the routing information and the inventory replenishment information. To facilitate our discussion, we define the following additional notations.

Additional sets and parameters

S: the complete set of feasible vehicle schedules for all planning periods. Define S_t where $t \in T$ is the set of all feasible vehicle schedules in period t. Therefore, we have $\bigcup_{t \in T} S_t = S$, and $S_{t_1} \cap S_{t_2} = \emptyset, \forall t_1, t_2 \in T$. Let S_{it} be the set of all vehicle schedules visiting customer i in period t.

c_s: the travel cost of schedule s, where $c_s = \sum_{ij \in S} c_{ij}$

b_{is}: the inventory delivered to customer $i \in M$ in schedule s

Additional variables

z_s: the binary variable indicating whether a vehicle schedule is selected in the solution. $z_s = 1$ if schedule s is selected and 0 otherwise.

u_{it}: the binary variable indicating whether a customer is visited in period t. $u_{it} = 1$ if customer i is visited in period t and 0 otherwise.

The schedule-based model for ML policy (SBM-ML) can be formulated as follows:

$$\min \sum_{s \in S} c_s z_s \tag{6.12}$$

$$s.t. \sum_{s \in S_t} z_s \leq |K| \quad \forall t \in T, \tag{6.13}$$

$$y_{it} = y_{it-1} + \sum_{s \in S_{it}} b_{is} z_s - d_{it} \quad \forall i \in M, \forall t \in T, \tag{6.14}$$

$$y_{it-1} + \sum_{s \in S_{it}} b_{is} z_s \leq B_i \quad \forall i \in M, \forall t \in T, \tag{6.15}$$

$$\sum_{s \in S_{it}} z_s = u_{it} \quad \forall i \in M, \forall t \in T, \tag{6.16}$$

$$z_s, u_{it} \in \{0,1\}, y_{it} \geq 0 \quad \forall i \in M, \forall t \in T, \forall s \in S. \tag{6.17}$$

The constraints in Equation 6.13 ensure that at most $|K|$ vehicle schedules can be selected in a period. The constraints in Equation 6.14 represent the inventory balance constraints for each customer. The constraints in Equation 6.15 ensure that inventory for customer i isless than or equal to its maximum inventory level. The constraints in Equation 6.16 ensure that at most one vehicle visits a customer in each period. The constraints in Equation 6.17 are nonnegative and binary constraints for variables.

The SBM-ML can be easily extended to OU policy by adding the following constraints.

$$y_{it} + d_{it} \geq B_i u_{it} \quad \forall i \in M, \forall t \in T. \tag{6.18}$$

The constraints in Equation 6.18 together with the constraints in Equation 6.14 ensure that the inventory at customer i is B_i after replenishment if u_{it} is 1.

6.4 Solution Methods

In this section, we first present a set of valid inequalities to tighten the linear relaxation of SBM. Then we develop a column generation algorithm to obtain the optimal LP relaxation of SBM. Finally, we propose a simple heuristic to obtain the IP solution to SBM efficiently.

6.4.1 Valid Inequality on the Number of Replenishments

Although the SBM improve the LP relaxation of BM to some extent, the gap between the optimal LP relaxation and the optimal IP solution of SBM is still large. This is because in the solution to the LP relaxation, the routing decisions are highly related to the inventory decision. Here, we present an example to show the reasons for the poor LP relaxation.

Consider a problem with a single vehicle, a single customer, and a single period. Customer i's daily demand $d_i = 5$; the vehicle's capacity $H = 25$; and we have three vehicle schedules s_1, s_2, and s_3, as shown in Table 6.1.

The optimal IP solution is $z_{s_1} = 1$ and the cost is 100. However, the optimal solution to the LP relaxation is $z_{s_2} = 0.2$ and $z_{s_3} = 0.8$, and the cost of the LP relaxation is 20. This is because the routing cost of the LP relaxation $\left(\dfrac{5}{25} \times c_{s_1}\right)$ is strongly affected by demand d_i in the SBM-LP relaxation. It is easy to see that the optimal LP relaxation is equal to $\dfrac{d_i}{H} \times 100$ for any d_i, and the IP–LP gap is $\dfrac{H - d_i}{H}$. When H is much larger than d_i, the LP relaxation of SBM provides little information on the optimal IP solution. In fact, this is also the case for BM.

It is not hard to see that the total number of visits to customer i in the complete planning horizon has to be greater or equal to $\left\lceil \dfrac{\sum_{t \in T} d_{it}}{\min\{B_i, H\}} \right\rceil$. Therefore, we can easily cut off the situation as shown in example above using the valid inequality that has been proposed in Adulyasak et al. (2014) and Coelho et al. (2014) for the IRP. That is,

$$\sum_{t \in T} u_{it} \geq \left\lceil \frac{\sum_{t \in T} d_{it}}{\min\{B_i, H\}} \right\rceil \quad \forall i \in M. \tag{6.19}$$

TABLE 6.1

Comparison between Optimal LP Relaxation and Optimal IP Solution to SBM

Schedule	Route	Inventory	Cost	LP Relaxation	Optimal IP
s_1	$O \to i \to O$	25	100	0.2	0
s_2	$O \to i \to O$	5	100	0	1
s_3	$O \to O$	0	0	0.8	0

The aforementioned bounds on the number of replenishments for any customer can be further improved as follows.

Proposition 6.1

For any customer with constant demand d_i, if $B_i \leq H$, then the number of replenishments has to be greater or equal to $\left\lceil \dfrac{T}{\left\lfloor \dfrac{B_i}{d_i} \right\rfloor} \right\rceil$. That is

$$\sum_{t \in T} u_{it} \geq \left\lceil \frac{T}{\left\lfloor \dfrac{B_i}{d_i} \right\rfloor} \right\rceil \quad \forall i \in M. \tag{6.20}$$

Proof: Because $B_i \leq H$, each time after replenishment, the maximum inventory is at most B_i. Therefore, it is easy to see that $\left\lfloor \dfrac{B_i}{d_i} \right\rfloor$ is the maximum number of periods that customer i can last without a replenishment. Then $\left\lceil \dfrac{T}{\left\lfloor \dfrac{B_i}{d_i} \right\rfloor} \right\rceil$ is the minimum number of replenishments needed for the complete planning horizon. EOF. ∎

Proposition 6.2

For any customer with constant demand d_i, if $B_i \leq H$, valid inequality in Equation 6.20 is stronger than the valid inequality in Equation 6.19.

Proof: Specifically, we have

$$\frac{T}{\lfloor B_i d_i \rfloor} = \frac{T \times d_i}{\left\lfloor \dfrac{B_i}{d_i} \right\rfloor \times d_i} \geq \frac{T \times d_i}{B_i} = \frac{\text{Total demand over } T}{B_i}.$$

Therefore, if we take the ceiling on both sides of this inequality, we have the following.

$$\left\lceil \frac{T}{\left\lceil \dfrac{B_i}{d_i} \right\rceil} \right\rceil \geq \left\lceil \frac{\text{Total demand over } T}{B_i} \right\rceil . \quad \text{EOF}$$

These bounds can be extended to nonconstant demand and the condition $B_i \leq H$ can also be relaxed. To find the bound on the minimum number of replenishments needed for any customer i, we first need to construct an inventory flow network $G_i(N_i, E_i)$ for each customer $i \in M$ as shown in Figure 6.1.

Each node $n \in N_i$ is indexed by two more parameters $t \in T$ and p, where $0 \leq p \leq B_i$ and p is integer. Define the set of nodes indexed by t as N_{it}. There are two types of arcs in the inventory flow network: consumption arcs and replenish arcs. A consumption arc e_{ipt} starts from node n_{ipt} and ends at node n_{iqt+1}, where $q = p - d_{it}$, representing that the inventory for customer i drops from p to q after one-period consumption. No consumption arc is constructed for node n_{ipt} if $p < d_{it}$. A replenishment arc e_{ipqt} starts from n_{ipt} and ends at node n_{iqt}, where $0 \leq p \leq q \leq B_i$, representing that the inventory for customer i is replenished from p to q. For an ML policy, p and q can be any integer as long as $q - p \leq H$ as shown in Figure 6.1a; and for an OU policy, q has to be equal to H and $p \geq \max\{B_i - H, 0\}$ as shown in Figure 6.1b. After the inventory flow network G_i is constructed, it is easy to see that any directed cycle in G_i can be viewed as a feasible replenishment plan for customer i. Then we define the following variables.

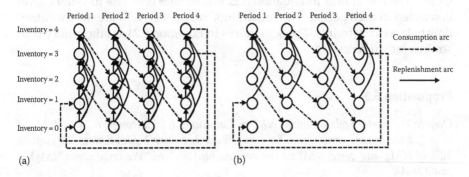

(a)
(b)

FIGURE 6.1
Inventory flow network containing four periods for a customer. The maximum inventory capacity of the customer is 4, and the demands are 2, 1, 2, and 3 for period 1, 2, 3, and 4, respectively. The vehicle capacity is 3. (a) Inventory flow network for ML policy. (b) Inventory flow network for OU policy.

Variables

r_{ipqt}: the binary variable indicating whether the replenishment arc e_{ipqt} is implemented. If $r_{ipqt} = 1$, the arc e_{ipqt} in period $t \in T$ is implemented and 0 otherwise.

w_{ipt}: the binary variable indicating whether the consumption arc e_{ipt} is used. If $w_{ipt} = 1$, the arc e_{ipt} is used and 0 otherwise.

Then we can solve the following network model (NM) for the minimum number of visits to each customer in the complete planning horizon.

$$\min \sum_{t \in T} \sum_{p=0}^{B_i-1} \sum_{q=p+1}^{B_i} r_{pqt} \tag{6.21}$$

$$s.t. \sum_{p=d_t}^{B_i} w_{pt} = 1 \quad \forall t \in T, \tag{6.22}$$

$$w_{i(p-d_{it-1})(t-1)} + \sum_{q=0}^{p-1} r_{qpt} = w_{pt} + \sum_{q=p+1}^{B_i} r_{pqt} \quad \forall t \in T, 0 \le p \le B_i, \tag{6.23}$$

$$w_{pt}, r_{pqt} \in \{0,1\} \quad \forall t \in T, 0 \le p < q \le B_i. \tag{6.24}$$

The objective in Equation 6.21 minimizes the total number of replenishments. The constraints in Equation 6.22 ensure that only one inventory level is selected in each period. The constraints in Equation 6.23 are the inventory flow balance constraints. The constraints in Equation 6.24 are the binary constraints for variables.

Proposition 6.3

Denote the partial relaxation of NM in Equations 6.21 through 6.24 as NM_w if we relax the binary constraints for r variables. Let $conv(NM_w)$ denote the convex hull of NM_w and $conv(NM)$ as the convex hull of NM. We have $conv(NM_w) = conv(NM)$.

Proof: For any feasible solution to NM_w, because w_{pt} variables have been fixed, the constraints in Equation 6.21 become void, and the constraints in Equation 6.23 form a network matrix (hence a totally unimodular matrix) with only r

variables, with a value on the right-hand side of 1, 0, or –1. Therefore, $\text{conv}(NM_w)$ has to be integral. Thus, we know $\text{conv}(NM_w) = \text{conv}(NM)$. EOF. ∎

Denote the objective value of NM_w for customer i as $\text{Obj}_i(NM_w)$, we have the following theorem.

Theorem 6.1

The number of replenishment for customer i has be to greater or equal to $\text{Obj}_i(MN_w)$. That is

$$\sum_{t \in T} u_{it} \geq \text{Obj}_i(NM_w) \quad \forall i \in M. \tag{6.25}$$

∎

Usually NM_w can be solved very efficiently because the majority of the constraints in NM_w are network constraints, and it contains only $t \times \min\{B_i, H\}$ number of binary variables.

We can also disaggregate the constraints in Equation 6.25 using the following inequality.

$$\sum_{t=t_1}^{t_1 + \tau_{it_1}} u_{it} \geq 1 \quad \forall i \in M, \forall t_1 \in T. \tag{6.26}$$

Here, τ_{it_1} is the minimum number of period starts from t_1 for customer i such that $\sum_{t=t_1}^{t_1 + \tau_{it_1}} d_{it} > B_i$. The inequality in Equation 6.26 ensures that there must be at least one visit to customer i from period t_1 to $t_1 + \tau_{it_1}$ because the total demand is more than the capacity of customer i.

Finally, we can also limit the total number of vehicle schedules used in the entire planning horizon as follows.

$$\sum_{t \in T} \sum_{s \in S_t} z_s \geq \left\lceil \frac{\sum_{t \in T} \sum_{i \in M} d_{it}}{H} \right\rceil. \tag{6.27}$$

The inequality in Equation 6.27 ensures there are enough vehicle schedules to carry the total demand for all customers during the entire planning horizon.

6.4.2 Nondominated Variables

It is obvious that SBM is a Dantzig–Wolfe decomposition of the BM. Like many Dantzig–Wolfe reformulations, SBM contains a very large number of z variables. The number of physical routes grow exponentially with the number of customers, and even for the same physical route, if the inventories delivered to each customer are different, they represent different z variables.

We can greatly reduce the size of S by excluding the dominated schedules in SBM. We say a schedule s_1 is dominated by route s_2, if $M_{s_1} \subseteq M_{s_2}$, $b_{is_1} \leq b_{is_2} \forall i \in M_{s_1}$, and $c_{l_1} \geq c_{l_2}$. Here, M_{s_1} and M_{s_2} are the sets of customers visited by s_1 and s_2 respectively. We can eliminate the dominated schedules in two stages. First, we know that for all vehicle routes (without considering inventory information) visiting the same set of customers, there exists one vehicle route that dominates all other vehicle routes with the minimum distance. Particularly, if the maximum number of customers that a vehicle can visit in a period is M_{veh}, the maximum number of routes is $\sum_{m=1}^{M_{veh}} \binom{|M|}{m}$.

Second, for the set of vehicle schedules constructed from a nondominated route, we can eliminate those that are not fully loaded. This is because for a vehicle schedule s_1 that is not fully loaded, there must exist a fully loaded vehicle schedule s_2 such that $b_{is_1} \leq b_{is_2} \forall i \in M_{s_1}$.

It is worth mentioning that if we erase all the dominated schedules, it might cause an infeasibility of SBM. For example, if we have a customer i with demand 5 and $B_i = 5$, and the vehicle capacity is 25. It is obvious that all the schedules, whose route is $o \rightarrow i \rightarrow o$, are dominated by the fully loaded vehicle with $b_{is} = 25$. If we erase all the dominated schedules such as the schedule $o \rightarrow i \rightarrow o$ with $b_{is} = 5$, the constraints in Equation 6.15 cannot be satisfied. To resolve this issue, we introduce a set of surplus variables w_{it}, which represents the unnecessary inventory if the inventory at a customer is more than its capacity. Therefore, the constraints in Equations 6.14 and 6.15 can be replaced as follows:

$$y_{it} = y_{it-1} + \sum_{s \in S_{it}} b_{is} z_s - w_{it} - d_{it} \quad \forall \in M, \forall t \in T, \tag{6.28}$$

$$y_{it-1} + \sum_{s \in S_{it}} b_{is} z_s - w_{it} \leq B_i \quad \forall i \in M, \forall t \in T, \tag{6.29}$$

$$w_{it} \geq 0 \quad \forall i \in M, \forall t \in T. \tag{6.30}$$

After solving SBM with the constraints in Equations 6.28 through 6.30, the actual inventory needed for each customer can be easily computed as $b_{is} - w_{it}$, $\forall i \in M_s, \forall s \in S_t, \forall t \in T$.

6.4.3 Column Generation

Because the number of schedules $|S|$ increases exponentially with the number of customers, it is impractical to enumerate all the nondominated routes, especially when the maximum number of customers that a vehicle can visit in a period is large. Therefore, we propose a column generation algorithm to obtain the optimal LP relaxation of SBM. To facilitate our discussion, we define the following dual variables.

Dual variables

- π: the positive dual variable associated with the constraints in Equation 6.27
- δ_t: the negative dual variable associated with the constraints in Equation 6.13
- α_{it}: the dual variable associated with the constraints in Equation 6.16
- β_{it}: the positive dual variable associated with the constraints in Equation 6.14 (because the = sign can be replaced by $\leq$)
- γ_{it}: the negative dual variable associated with the constraints in Equation 6.15

The reduced cost of a schedule c'_s in period t can be written as follows:

$$
\begin{aligned}
c'_s &= c_s - \sum_{i \in M_s} \alpha_{it} - \sum_{i \in M_s} b_{is}\beta_{it} - \sum_{i \in M_s} b_{is}\gamma_{it} - \delta_t - \pi \\
&= \sum_{ij \in l} c_{ij} - \sum_{i \in M_s} \alpha_{it} + (b_{is}(\beta_{it} + \gamma_{it})) - \delta_t - \pi \\
&= \sum_{ij \in l} (c_{ij} - \alpha_{it} - b_i(\beta_{it} + \gamma_{it})) - \delta_t - \pi \quad \forall s \in S_t, \forall t \in T.
\end{aligned}
\tag{6.31}
$$

Therefore, if we can find any schedule s such that the reduced cost is less than 0, we should add it to SBM to improve the LP relaxation. If there is no schedule such that the reduced cost is less than 0, we know the current LP relaxation is optimal. From Equation 6.31, we can easily see that the pricing subproblem in period t can be formulated as follows:

$$
\min \sum_{i,j \in M} (c_{ij} - \beta_{it} - (\alpha_{it} + \gamma_{it})b_i)x_{ij}
\tag{6.32}
$$

$$
s.t. \sum_{j \in M} x_{m_0 j} = 1
\tag{6.33}
$$

$$\sum_{j \in M} x_{jm_o} = 1 \tag{6.34}$$

$$\sum_{j \in M} x_{ij} - \sum_{j \in M} x_{ji} = 0 \quad \forall i \in M, \tag{6.35}$$

$$b_i \le \sum_{j \in M} \min\{H, B_i\} x_{ij} \quad \forall i \in M, \tag{6.36}$$

$$\sum_{i \in M} b_i \le H \tag{6.37}$$

$$x_{ij} \in \{0,1\}, \, b_i \ge 0 \quad \forall i, j \in M. \tag{6.38}$$

Here, we assume $\alpha_{ot} = \beta_{ot} = 0$ for the consistency. The objective function in Equation 6.32 minimizes the reduced cost of the schedule. The constraints in Equations 6.33 through 6.35 are the flow balance constraints for the route. The constraints in Equation 6.36 ensure the delivered inventory b_i can be positive only if customer i is visited. The constraints in Equation 6.37 are the capacity constraints. Although the objective function in Equation 6.32 is quadratic, it can be linearized as follows:

$$\min \sum_{i, j \in M} (c_{ij} - \alpha_{it}) x_{ij} - \sum_{i \in M} (\beta_{it} + \gamma_{it}) b_i \tag{6.39}$$

Here, we provide two simple heuristics to the pricing subproblem for SBM. In the first heuristic, we first discard the profit contributed by $(\beta_{it} + \gamma_{it}) b_i$, and then the problem becomes the shortest path problem. Because the value of $c_{ij} - \alpha_{it}$ can be positive or negative, we apply the well-known label correction algorithm, Bellman–Ford algorithm, to obtain the optimal solution to the shortest path problem. Then we can sort all the customers in the shortest path by $\beta_{it} + \gamma_{it}$ in descending order. We first ensure that each customer has the minimum inventory delivered, that is, $b_i = 1$, $\forall i$. Then we try to maximally allocate the remaining capacity to the customer with the largest $\beta_{it} + \gamma_{it}$. If there is still remaining capacity left, we repeat the procedure to the customer with the second largest $\beta_{it} + \gamma_{it}$. The procedure continues until there is no vehicle capacity left. In the second heuristic, we sort all the customers by $\beta_{it} + \gamma_{it}$ in descending order, and we try to allocate the maximum capacity to the customers with higher $\beta_{it} + \gamma_{it}$ until there is no capacity left. We then construct the shortest path to the selected customers and compute the final reduced cost.

In every iteration of the column generation, after solving the subproblem by two proposed heuristics, if the reduced cost for some schedule is negative, we add it to the restricted master problem without solving the pricing sub-problem model in Equations 6.32 through 6.38. Otherwise, we need to use the subproblem model in Equations 6.32 through 6.38 to prove the optimality of the LP relaxation. For each column generation iteration, we need to solve $|K| \times |T|$ subproblems.

6.4.4 Implementation Issues

We can always use branch-and-price to obtain the optimal IP solution for SBM. However, in our study, instead of using branch-and-price, we solve a restricted IP with only the schedules generated in the column generation. To speed up the computation, we first modify the constraints in Equation 6.13 as follows:

$$v_t - \sum_{s \in S_t} z_s \geq 0 \quad \forall t \in T, \tag{6.40}$$

$$0 \leq v_t \leq H, \text{ and is integer} \quad \forall t \in T. \tag{6.41}$$

Here, v_t represents the number of vehicles used in each period. Then, in the branch-and-bound tree, we always branch on v_t variables first, followed by u_{it} variables, and lastly on z_s variables. This is because in each period, the number of vehicles used is a higher level decision than which customers to visit, and u_{it} (the set of customers to visit) leads to the decision on the exact vehicle schedules. The above branching rule can be achieved using the CPXcopyorder function. Also, when there is a fraction variable v_t or u_{it}, we always first search the floor of that variable because we want to reduce the number of vehicles used in each period (with respect to v_t) and the total number of visits to all customers (with respect to u_{it}) because intuitively it might reduce the total number of vehicles and replenishment to customers. When we solve the SBM, we stop the CPLEX branch-and-bound process if there is no improvement on the best integer solution within 1800 seconds.

6.5 Model Extension

In this section, we discuss several possible model extensions that could be useful in real-life applications.

6.5.1 Multiple Depots

Assume there are multiple depots denoted by set O, and a vehicle can depart from and arrive at different depots during its route. Therefore, when we generate vehicle schedules, a route can start and end at different depots. Also, we need to ensure that the number of vehicles that arrived at a depot in the previous period must be equal to the number of vehicles that depart from the same depot in the next period, so that the vehicle schedule is repeatable. This can be formulated as follows:

$$\sum_{s \in S_{ot}^+} z_s = \sum_{s \in S_{ot+1}^-} z_s \quad \forall o \in O, \ \forall t \in T. \tag{6.42}$$

Here, S_{ot}^+ and S_{ot}^- are the sets of vehicle schedules arriving at and departing from depot $o \in O$ in period t respectively.

6.5.2 Depot Capacity

If the inventory provided by the depot in each period is limited and denoted by $Cap_o, \forall_o \in O$, we can formulate the depot capacity constraints as follows:

$$\sum_{s \in S_{ot}} b_s z_s \leq Cap_o \quad \forall o \in O, \ \forall t \in T. \tag{6.43}$$

Here, b_s is defined as the total delivery amount of vehicle schedule s, so $b_s = \sum_{i \in s} b_{is}$.

6.5.3 Multiple Fleets

In many real-life situations, more than one fleet is available. Therefore, it is natural to extend the proposed model for the multiple-fleet situation. This can be easily done by introducing a new set of available fleets, F, and increasing the dimensionality of variables with a fleet index. Define K_f as the set of vehicle available in fleet $f \in F$. Define S_{fit} as the set of vehicle schedules for fleet f that visit customer i in period t. The MIP formulation for multifleet CIRP with multiple-depot and depot capacity is given by

$$\min \sum_{f \in F} \sum_{s \in S_f} c_s z_s \tag{6.44}$$

$$s.t. \sum_{s \in S_{ft}} z_s \leq |K_f| \quad \forall f \in F, \forall t \in T, \tag{6.45}$$

$$y_{it} = y_{it-1} + \sum_{f \in F} \sum_{s \in S_{fit}} b_{is} z_s - d_{it} \quad \forall i \in M, \forall t \in T, \tag{6.46}$$

$$y_{it-1} + \sum_{f \in F} \sum_{s \in S_{fit}} b_{is} z_s \leq B_i \quad \forall i \in M, \forall t \in T, \tag{6.47}$$

$$\sum_{t \in T} \sum_{f \in F} \sum_{s \in S_{fit}} z_s \geq \left\lceil \sum_{t \in T} \sum_{i \in M} d_{it} \max_{f \in F} H_f \right\rceil, \tag{6.48}$$

$$\sum_{s \in S_{fot}^+} z_s = \sum_{s \in S_{fot+1}^-} z_s \quad \forall f \in F, \forall o \in O, \forall t \in T, \tag{6.49}$$

$$\sum_{f \in F} \sum_{s \in S_{fot}} b_s z_s \leq \mathrm{Cap}_o \quad \forall o \in O, \forall t \in T. \tag{6.50}$$

Equations 6.25 and 6.26,

$$z_s, u_{it} \in \{0,1\}, 0 \leq y_{it} \leq B_i \quad \forall f \in F, \forall i \in M, \forall t \in T, \forall s \in S_{fit}. \tag{6.51}$$

6.5.4 Finite Period Inventory Routing Problem

In fact, SBM can also be extended to model the finite period IRP because CIRP and IRP are closely related. To facilitate our discussion, we denote I_{it_0} as the initial inventory for customer i. Then we just a need to discard the inventory balance constraints in Equation 6.14 for $t = |T|$, and set $y_{it_0} = I_{it_0}$.

To compute the minimum number of replenishments in Equation 6.25, we only build the inventory consumption arcs and replenishment arcs corresponding to inventory level I_{it_0} for customer i in period t_0 as shown in Figure 6.2.

All the inventory consumption arcs and replenishment arcs terminate at the nodes in the last period of the planning horizon (hence the schedule is no longer cyclic). Then the proposed SSM model can be easily adapted to the new inventory flow network.

Finally, it is worth mentioning that in general CIRP is more difficult than the finite period IRP because of the following two points. First, the solution space of CIRP is larger than IRP because in CIRP there is no initial inventory. As a result, any combination of the initial inventories to all customers may appear in the optimal solution to CIRP. Second, many known results

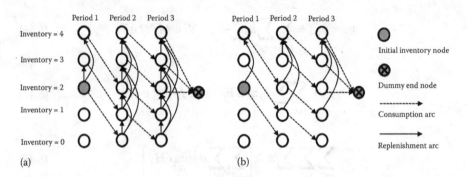

FIGURE 6.2
Inventory flow network for a three-period inventory routing problem. The initial inventory level is 2. The maximum inventory capacity of the customer is 4, and the demands are 2, 1, and 2 for period 1, 2, and 3, respectively. The vehicle capacity is 3. (a) Inventory flow network for ML policy. (b) Inventory flow network for OU policy.

for IRP can no longer be applied to CIRP. For example, Archetti et al. (2007, 2011) and Adulyasak et al. (2014) proposed a very effective valid inequality as follows:

$$\sum_{t=t_0}^{t'}\sum_{i \in M_s} u_{it} \geq \left\lceil \frac{\sum_{i \in M_s} \max\left\{0, \sum_{t=t_0}^{t'} d_{it} - I_{it_0}\right\}}{H} \right\rceil \quad \forall i \in M, \ \forall M_s \subseteq M. \qquad (6.52)$$

This valid inequality ensures that the total number of vehicles entering and leaving the set of customers M_s from period t_0 to t' must be sufficient to carry the demands in these periods. However, we cannot adapt this valid inequality in CIRP because we cannot substitute I_{it_0} by $\sum_{p=d_{it_0}}^{B_i} p w_{ipt_0}$; otherwise the inequality becomes nonlinear.

6.6 Computational Study

In this section, we present the computational results of SBM. We first introduce five real-life test cases used. Then we present the computational solution for small and mid-size test cases by enumerating all the possible schedules. We also present the solution to all test cases using the column generation algorithm. Finally, we analyze the reduced cost of the schedules in the best IP solution of SBM.

6.6.1 Details of the Test Instances

Our test instances are provided by an industrial gases manufacturer. We have a single fleet of 23-ton vehicles. The inventory capacities at customers range from 7 tons to 30 tons. The inventory cost at any customer is 0. The vehicle speed is assume to be 50 kilometers per hour constantly, a vehicle can work at most 12 hours every day, and a vehicle cannot serve more than five customers per day. We assume the unload time at each customer is 1 hour, and the minimum delivery amount to any customer has to be no less than 2 tons. We try to build a weekly cyclic inventory routing plan, in which each day is considered as a single period (so $|T| = 7$). The detailed information on each test case is shown in Table 6.2.

6.6.2 Solution by Enumerating All Schedules

We first solve the first three instances with SBM by enumerating all the possible vehicle schedules. We set the computational time to be 4 hours for all test cases. The results are shown in Table 6.3.

We can see from Table 6.3 that the SBM can obtain optimal solutions to five out of six small and mid-size test cases. The only nonoptimally solved test case 2-ML has an optimality gap of 3.3%. However, the best solution to 2-ML

TABLE 6.2

Detailed Information about the Five Real-Life Test Instances

Instances	Number of Customers	Number of Vehicles	Number of Depots	Average Daily Demand
1	5	2	1	35
2	0	3	1	38
3	20	5	1	49
4	47	10	1	185
5	67	15	1	235

TABLE 6.3

Computational Results by Enumerating All of the Schedules for Small and Mid-size Test Cases

Test Instances	Policy	Number of Schedules	LP Value	LP Time	IP Value	IP Time	CPLEX Optimal Gap (%)	Root Node IP–LP Gap (%)
1	ML	46,795	4340	1	4365	61	0.0	0.6
	OU	46,795	4340	1	4642	120	0.0	6.5
2	ML	148,820	4973	8	5209	14,400	3.3	4.5
	OU	148,820	4973	8	5318	388	0.0	6.3
3	ML	513,513	8443	12	8651	374	0.0	2.4
	OU	513,513	8443	11	8895	1416	0.0	5.1

TABLE 6.4

Computational Result Using Column Generation

Test Instances	Policy	Number of Iterations	Number of Schedules	LP Value	LP Time	IP Value	IP Time	IP–LP Gap (%)	Enumeration Solution (%)
1	ML	3	3388	4340	1	4370	11	0.7	0.1
	OU	3	3388	4340	1	4642	37	6.5	0.0
2	ML	3	8701	4973	2	5209	1816	4.5	0.0
	OU	3	8701	4973	2	5318	244	6.3	0.0
3	ML	3	2940	8443	3	8651	10	2.6	0.0
	OU	3	2968	8443	3	8927	17	2.8	0.4
4	ML	26	108,556	17,077	802	18,153	14,400	5.9	–
	OU	24	136,440	17,077	663	19,187	14,400	11.1	–
5	ML	34	232,357	22,107	1261	23,522	14,400	6.0	–
	OU	39	291,472	22,107	1452	24,397	14,400	9.4	–

is found in the very early stage of the branch-and-bound process. For other test cases, the solution time of the OU policy is slightly longer than that of the ML policy. This might be because the number of constraints in the OU policy is higher than that of the ML policy. Finally, it is important to note that the LP relaxation of SBM is very good. The average gap between the best IP solution and the LP relaxation is 3.3% for the ML policy and 6.2% for the OU policy.

6.6.3 Column Generation Results

We also solve the LP relaxation of SBM using column generation. Then we solve a restricted IP problem with only the schedules generated in the column generation procedure. We show the computational results in Table 6.4.

As we can see from Table 6.4, we can obtain very good solutions using column generation. For small and mid-size test cases, the best integer solutions obtained are within a 0.1% gap on average compared with the solutions obtained using enumeration methods. The computational times for test cases 1–3 are reduced drastically. However, the computational time for large test cases 4 and 5 is still 4 hours. The IP–LP gaps for test cases 4 and 5 are about 6% for the ML policy and 10% for the OU policy, respectively.

6.6.4 Schedule Quality in the Best IP Solutions

Finally, we also report the LP reduced cost of the vehicle schedule in the best IP solutions.

As we can see from Table 6.5, for the ML policy, the reduced cost of 80% of the schedules is less than 5. For the OU policy, the reduced cost of more than

TABLE 6.5

Vehicle Schedule Quality in the Best IP Solutions

Test Instances	Policy	Total Schedules	Value of the Reduced Cost c_s'					
			0	(0, 5]	(5, 20]	(20,50]	(50, +∞)	Average
1	ML	1	6 (55)	3 (27)	2 (18)	0 (0)	0 (0)	1.60
	OU	12	4 (33)	6 (50)	2 (17)	0 (0)	0 (0)	2.85
2	ML	17	13 (76)	4 (24)	0 (0)	0 (0)	0 (0)	0.42
	OU	17	11 (65)	3 (18)	3 (18)	0 (0)	0 (0)	1.54
3	ML	23	17 (74)	2 (9)	2 (9)	1 (4)	1 (4)	6.24
	OU	23	16 (70)	1 (4)	0 (0)	5 (22)	1 (4)	14.5
4	ML	61	27 (44)	19 (31)	11 (18)	4 (7)	0 (0)	4.22
	OU	63	21 (33)	15 (24)	17 (27)	8 (13)	2 (3)	10.21
5	ML	76	28 (37)	20 (26)	25 (33)	3 (4)	0 (0)	4.54
	OU	78	22 (28)	26 (33)	20 (26)	9 (11)	1 (1)	7.34
Average	ML	–	57%	23%	16%	3%	1%	3.4
	OU	–	46%	26%	18%	9%	2%	7.3

Note: Numbers in parentheses are percents.

70% of the schedules is less than 5. The average reduced cost of all schedules is only 3.4 for the ML policy and 7.3 for the OU policy. For the ML policy, there are only 4% of the schedules whose reduced cost is more than 20, and this number is 11% for the OU policy. From this table, we can see that the reduced cost can be viewed as a reasonable measurement for the quality of a schedule.

6.7 Conclusions

In this chapter, we proposed a schedule-based model (SBM) for the cyclic inventory routing problem. This model can be considered a Dantzig–Wolfe reformulation of the widely used arc-based model for the IRP. To solve SBM efficiently, we developed a column generation algorithm, in which only attractive vehicle schedules are generated. We also proposed a group of new valid inequalities to tighten the bounds of SBM. Our computational results on five real-life test cases show the LP relaxation of SBM is quite tight. SBM can obtain near-optimal solutions to very large real-life test cases within a reasonable time, and IP–LP gaps of SBM are within 5% and 7% for the ML policy and the OU policy on average for all test cases, respectively.

There are several possible future research directions in this research. First, we can further develop new valid inequalities to tighten the bound of the LP relaxation of SBM. Second, as we discussed in Section 6.3.8, we can test the proposed model on the finite period inventory routing problem, especially on the well-studied test cases provided in Archetti et al. (2007, 2012). Third, we could also extend the current deterministic SBM model to the stochastic cyclic inventory routing problem with stochastic demand.

Acknowledgments

Zhe Liang was supported by the National Science Foundation of China (Grant 71422003 and Grant 71201003).

References

Adelman, D. A price-directed approach to stochastic inventory routing. *Operations Research*, 52 (4): 499–514, 2004.

Adulyasak, Y., J.-F. Cordeau, and R. Jans. Formulation and branch-and-cut algorithms for multivehicle production and inventory routing problem. *INFORMS Journal on Computing*, 26(1): 103–120, 2014.

Aghezzaf, E.-H., B. Raa, and H. V. Landeghem. Modeling inventory routing problems in supply chain of high consumption products. *European Journal of Operational Research*, 169(3): 1048–1063, 2006.

Aghezzaf, E.-H., Y. Zhong, B. Raa, and M. Mateo. Analysis of the single-vehicle cyclic inventory routing problem. *International Journal of Systems Science*, 43(11): 2040–2049, 2012.

Andersson, H., A. Hoff, M. Christiansen, G. Hasle, and A. Lokketangen. Industrial aspects and literature survey: Combined inventory management and routing. *Computers & Operations Research*, 37(9): 1115–1136, 2010.

Anily, S. and F. Federgruen. One warehouse multiple retailer systems with vehicle routing costs. *Management Science*, 36(1): 92–114, 1990.

Archetti, C., L. Bertazzi, G. Laporte, and M. G. Speranze. A branch-and-cut algorithm for a vendor-managed inventory-routing problem. *Transportation Science*, 41(3): 382–391, 2007.

Archetti, C., L. Bertazzi, G. Paletta, and M. G. Speranze. Analysis of the maximum level policy in a production-distribution system. *Computers and Operations Research*, 38: 1731–1746, 2011.

Archetti, C., L. Bertazzi, A. Hertz, and M. G. Speranze. A hybrid heuristic for an inventory rotuing problem. *INFORMS Journal on Computing*, 24(1): 101–116, 2012.

Baldacci, R., N. Christofides, and A. Mingozzi. An exact algorithm for vehicle routing problem based on the set partitioning formulation with additional cuts. *Operations Research*, 59: 1269–1283, 2011.

Chan, L. M. A., M. G. Speranza, and L. Bertazzi. Asymptotic analysis of periodic policies for the inventory routing problem. *Naval Research Logisitics*, 60: 525–540, 2013.

Coelho, L. C., J.-F. Cordeau, and G. Laporte. Consistency in multi-vehicle inventory-routing. *Transportation Research Part C*, 24: 270–287, 2012.

Coelho, L. C., J.-F. Cordeau, and G. Laporte. Thirty years of inventory routing. *Transportation Science*, 480(1): 1–19, 2014.

Ekici, A., O. O. Ozener, and G. Kuyzu. Cyclic delivery schedules for an inventory routing problem. *Transportation Science*, Forthcoming, 1–13: 2015 [epub ahead of print].

Engineer, F. G., K. C. Furman, G. L. Nembauser, M. W. Savelsbergh, and J.-H. Song. A branch-and-price-and-cut algorithm for single-product maritime inventory routing. *Operations Research*, 60(1): 106–122, 2012.

Fukasawa, R., H. Longo, J. Lysgaard, M. P. De Aragao, M. Reis, E. Uchoa, and R. F. Werneck. Robust branch-and-cut-and-price for the capacitated vehicle routing problem. *Mathematical Programming*, 106: 491–511, 2006.

Gallego, G. and D. Simchi-Levi. On the effectiveness of direct shipping strategy for the one-warehouse multi-retailer r-system. *Management Science*, 36(2): 240–243, 1990.

Gronhaug, R., M. Christiansen, G. Desaulniers, and J. Desrosiers. A branch-and-price method for a liquefied natural gas inventory routing problem. *Transportation Science*, 44(3): 400–415, 2010.

Hvattum, L. M., A. Lokketangen, and G. Laporte. Scenario tree-based heuristics for stochastic inventory-routing problems. *INFORMS Journal on Computing*, 21(2): 268–285, 2009.

Kleywegt, A. J., V. S. Nori, and M. W. P. Savelsbergh. The stochastic inventory routing problem with direct deliveries. *Transportation Science*, 36(1): 94–118, 2002.

Kleywegt, A. J., V. S. Nori, and M. W. P. Savelsbergh. Dynamic programming approximations for a stochastic inventory routing problem. *Transportation Science*, 38(1): 42–70, 2004.

Pecin, D., A. Pessoa, M. Poggi, and E. Uchoa. Improved branch-and-cut-and-price for capacitated vehicle routing. In *Integer Programming and Combinatorial Optimization* (pp. 393–403). Lecture Notes in Computer Science, Vol. 8494. Berlin: Springer Science+Business Media, 2014.

Raa, B. and E.-H. Aghezzaf. Designing distribution patterns for long-term inventory routing with constant demand rates. *International Journal of Production Economics*, 112: 255–263, 2008.

Raa, B. and E.-H. Aghezzaf. A practical solution approach for the cyclic inventory routing problem. *European Journal of Operational Research*, 192: 429–441, 2009.

Solyli, O. and H. Sural. A branch-and-cut algorithm using a strong formulation and an a priori tour-based heuristic for an inventory-routing problem. *Transportation Science*, 45(3): 335–345, 2011.

Solyli, O., J.-F. Cordeau, and G. Laporte. Robust inventory routing under demand uncertainty. *Transportation Science*, 46(3): 327–340, 2012.

Toth, P. and D. Vigo. *Vehicle Routing Problems, Methods and Applications*, 2nd ed. Philadelphia: Society for Industrial and Applied Mathematics and Mathematical Optimization Society, 2014.

Vansteenwegen, P. and M. Mateo. An iterated local search algorithm for the single-vehicle cyclic inventory routing problem. *European Journal of Operational Research*, 237: 802–813, 2014.

Zhao, Q.-H., S. Chen, and C.-X. Zang. Model and algorithm for inventory routing decision in a three-echelon logisitics system. *European Journal of Operational Research*, 191: 623–635, 2008.

7

An Application of an Inventory Model for Production Planning

Paveena Chaovalitwongse, Pakpoom Rungchawalnon,
and Kwankeaw Meesuptaweekoon

CONTENTS

ABSTRACT   A good alignment between customer demand (in terms of
quantity and timing) and production orders results from an effective and
efficient production plan. This chapter concerns the production planning
of a production line in a finishing process of rolled tissues at a case study
manufacturer. The case study is making rolled tissues under a make-to-
stock manufacturing environment, which is a normal practice for commodity products. Thus, its production plan can ensure that the customer service
level requirements will be achieved at a "just-enough" level of inventory.
Currently, the case study cannot fulfill customer demand for all items at the
required service level. In addition, it is not clear how much inventory is just
enough for them. Thus, this chapter aims to improve the current production planning method to serve customers better with an appropriate level of
inventory. The simple reorder point fixed order quantity inventory model is
introduced in a new production planning method that comprises two components: establishing inventory policy and determining production orders.
The proposed method is tested with the actual demand data and then compared with the current method. The results show that with the proposed

method all items can pass the service level requirements. In addition, there are savings from inventory and setup reduction of approximately 37% and 27% respectively.

KEY WORDS:　*inventory model, make-to-stock, production plan, reorder point.*

7.1 Introduction

The case study company is a large household product manufacturer. One of its lines of products is rolled tissues, which include bathroom tissues and kitchen towels. These products vary in the quality of tissue paper, types of wrapping sheet, package size, and length of paper per roll.

This chapter focuses on the finishing process of rolled tissue making in which jumbo tissue rolls are processed into rolls of usable size. The finishing process encompasses embossing, winding to length, cutting to size, wrapping, and boxing (see Figure 7.1). The jumbo tissue rolls are produced from a preparatory process. According to the case study policy, jumbo rolls must be maintained for the finishing process at all times. Unlike jumbo tissue rolls, plastic paper and cardboard boxes are procured from outside sources. The procurement plan for these two materials must be align with the finishing process production plan.

A single continuous line is used in this finishing process. The production line is an automatic line and can be adjusted to process various types of finished rolled tissue products. In other words, multiple products share the same production line.

The production line capacity is determined by the number of cases that can be produced per day. Because the processing times are different among finished goods, the capacity for each finished good can be unequal. There will be a production line setup once a new product is produced. Thus, one setup occurs when a new lot size is produced. It is noted that the major loss from the setup is material loss. The setup time loss between lot sizes is only

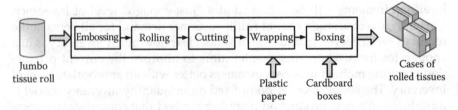

FIGURE 7.1
A finishing process of rolled tissues.

slightly due to the technical specification of the production line. Therefore, it allows us to neglect this time loss when determining a production plan.

The main customers for rolled tissues are large-scale retailers. The case study's customers send their purchase order via Electronic Data Interchange (EDI) and require an order fulfillment time of 1 day. Moreover, the service agreement has a 99% case fill rate. If the case study fails to fulfill customer orders, there will be a high penalty cost.

Currently, the case study plans its production under a make-to-stock environment. Production planning is determined weekly in a 4-week advance period by using monthly forecasts from the Sales Department. After determination of a production plan, the Material Requirement Planning (MRP) for plastic paper and cardboard boxes is calculated. With a highly fluctuating demand pattern, the forecast is rarely accurate. As a result, the current production plan does not meet the actual demand and leads to a poor fill rate. Thus, one of the key issues in satisfying customer demand under a make-to-stock environment is good production planning. In other words, the production planning must be able to align the production plan with customer demand. The effectiveness of production planning can be measured by service level, average inventory, and the number of machine setups.

This chapter is organized as follows. In Section 7.2 we review some of the pertinent literature. In Section 7.3, we study and analyze the as-is production planning method. The proposed method is described in Section 7.4. The proposed method evaluation and results are reported in Section 7.5. Finally, in Section 7.6 we make some concluding remarks.

7.2 Literature Review

To achieve customer satisfaction, demand fulfillment is always challenging in many businesses. Maintaining customer loyalty might require a good management plan at a great cost because stationary demand rarely exists in the real world (Agrawal et al., 2009). In many industries, products have a slight differentiation among brands (Rego and Mesquita, 2015). A high inventory level might provide better customer service but it may entail a great inventory cost. However, low inventory may pose a risk of lowering the customer service level and lost sales due to the shortage of goods (Gruen and Corsten, 2003). Therefore, many businesses have tried to study demand patterns to forecast demands in the future to manage production plans, material requirement plans, and inventory of finished goods to be replenished on time.

Demand variability is affected by many components, for example, unstable demands of end customers and the results of the bullwhip effect in the supply chain. The literature review shows that several studies are focusing on different aspects of demand forecasting and inventory control. However, most of them focus only on how to deliver the best model to forecast demands with minimum error by considering external factors.

On the supply side, most models assume that the Production Department is ready to fulfill orders according to the inventory policy setting. In such a setting, an order is placed based on the demand prediction coming from the forecasting model, but the prediction model does not consider the nature and constraints of production lines. Most research on the integrated production–inventory model consider the collaboration between the echelons. Vendor-managed inventory (VMI) is a particularly interesting approach to solve this problem and has been progressively applied in several companies and research studies (Zavanella and Zanoni, 2006). However, they consider only a production rate in terms of a constraint or an assumption of the production part and pay attention to a synchronized inventory policy and the collaborative strategy between parties (Boyaci and Gallego, 2002; Hoque and Goyal, 2006; Sarmah et al., 2006).

If considering the other type of production–inventory system, the Capacitated Production–Inventory System and Production–Priority Policy might have more correspondence with our work. They are particularly relevant to producing multiple products for which demand is nonstationary and the products share a finite-capacitated resource problem that has been studied and solutions proposed in some interesting work. A base stock policy is proposed to solve this problem by using an optimal method and a heuristic to compute the levels of heterogeneous and homogeneous products (DeCroix and Arreola-Risa, 1998). However, the assumption concerning production is typically the same as in other work as well as in the work of Balkhi (2009), who presented the optimal stopping and restarting production times for each produced item, assuming production capacity is always available to produce goods at the production rate of each period of time the same as other Capacitated Production–Inventory works (Özer and Wei, 2004).

Nevertheless, production and inventory management problems with a single resource constraint have received a low level of attention in the literature (Bretthauera et al., 2006). Reorder cycle times that are independent for each item carried in inventory is the ordinary policy to apply in this problem. Because there is a chance that many products may eventually be ordered that need to be produced at the same time, the approach in these works is to manage the situation to satisfy constraints and also the objective functions. The optimal Lagrange multiplier and its improvement (Maloney and Klein, 1993) are proposed as one of two types of solution approaches. The other type is an improvement approach for considering order quantity (Page and Paul, 1976) and cycle times of each item in the system. However, the computational effort is not appropriate for implementation.

The concept of inventory acting as a buffer, to absorb increases or decreases in demand while production remains relatively steady, is also proposed by Buffa and Miller (1979). Also, in the real case, production is not stable because of the constraint and capacity. Therefore, in this work, we represent the simple method to deal with the variability of demand while production has a strict constraint in terms of capacity and availability of a machine that is shared among multiple products.

7.3 As-Is Study and Analysis

7.3.1 Current Production Planning Process

The current production plan is driven by the monthly demand forecast or 4-week time frame that has been agreed on by the Sales and Production Departments. The 4-week demand forecast for each finished goods item is then distributed to weekly demand by the production planner. It is noted that there is no concrete rule or guideline on how it should be done. Once the weekly forecast has been acknowledged, the planner decides how much of each product should be produced each week according to production line capacity and forecast demand (no backlog is allowed in planning). Then the planner must determine the details of production orders that indicate what product/when to start–finish/how much (order quantity or lot size) to produce within the week. After the production plan and orders have been issued, the planning of raw materials is determined by MRP. It should be observed that because of the lead time of procuring raw materials, the production orders for the first 2 weeks of the 4-week plan must consider the raw material availability. The production plan is officially revised weekly. Figure 7.2 represents the current process of production planning by the ICOM (input/output/constraints/mechanism) model.

7.3.2 Current Performance

The performance of the current planning method in 2010 is shown in Figure 7.3. It indicates that only 3 of 10 finished goods (FG) items have a service level greater than 99% or achieve a satisfactory customer service level. In addition, it is not clear whether the inventory level is appropriate or not. For FG02 and FG03, they can meet a 99% service level with fewer than 15 days of inventory sales. However, FG09 needs to hold inventory over 30 days of sales to a satisfactory customer service level. For other items, it may seem that the holding inventory is too low to achieve a satisfactory customer service level.

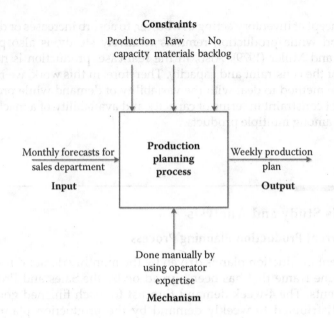

FIGURE 7.2
Overview of current production planning process.

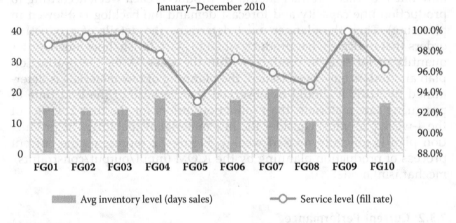

FIGURE 7.3
Production planning performance indicators.

Besides the aforementioned performance indicators, the frequency of the production line setup is also considered. With the current planning method, emergency changes always occur in the production plan if the raw materials are available, largely because of using inaccurate forecast data to determine a production plan that cannot align with incoming orders. Thus, at this point, the current planning method is not good enough to match the customer demand with the given production plan.

7.3.3 An As-Is Method Analysis

In planning production, the major concern is aligning customer demands with production orders. To obtain better alignment, one should first understand the characteristics of demand and production (see Table 7.1).

According to the current planning method, the forecast used does not give a good estimate of the actual customer demand, which has a 1-day order fulfillment time and high daily fluctuation. In addition, it is not likely to get a good daily demand forecast for commodity products such as rolled tissues. Therefore, production planning should not be based mainly on forecasting.

Currently, there is no decision rule or guideline for a production planner to follow to ensure that production orders are aligned with incoming demands. As a result, the planner faces emergency setup changes. According to the case study planning procedure, the production plan should be reviewed once a week. As for current operations, emergency production changes occur almost every day of the week. As a result, the initial plan has been overwritten and never used. It is also difficult for planning raw material procurement. Thus, the new production planning should include a decision guideline to react rationally to actual demands.

Considering production characteristics, the production line allows a new setup with little time loss. Thus there is no need for a long-term planning period to minimize the setup time loss between product changes. Because the setup

TABLE 7.1

Characteristics of Demand and Production

Demand Characteristics	Production Characteristics
1. Daily demands of each individual item are highly fluctuating and cannot be accurately forecast.	1. Production is performed on a single continuous line that must be shared among 10 products.
2. The total quantity of demands for any given day can exceed the production capacity.	2. The production line is continuous and fully automated; thus job preemption is not allowed.
3. Annual demands are more stable and predictable.	3. The production line is capacitated.
4. The order fulfillment time is 1 day.	4. There is a setup every time when changing products on the production line.
5. The service level must exceed 99%.	5. The setup results in loss of material (tissue paper) and time. However, the time loss is very small and can be neglected.
6. The penalty cost when failing to fulfill orders is high.	6. There are differences in production capacity among products.
	7. The production plan indicates what/when/how much (lot size) to make in each lot size or every time a production line setup occurs.
	8. When producing more than enough, an inventory holding cost is incurred.
	9. The cost of external raw materials (plastic wrap and cardboard boxes) is very low compared to the penalty cost from customers.

loss accounts mainly for material loss, the key consideration here is to determine the appropriate lot size for each product setup on the production line.

Raw material planning should not be done by MRP because the production plan is not static and subject to change with highly fluctuating daily demands. In addition, the raw material costs are much lower than the penalty cost imposed by the customers. It is better to have raw materials available to ensure a quick response to the dynamics of demands.

In sum, the current method can be improved in several aspects: (1) The production plan should not be based mainly on forecasts, (2) there is a need for a decision guideline, (3) the economic production lot size should be considered for each production setup, and (4) raw materials should be ready for uncertain production orders due to the demands of the dynamics.

7.4 Proposed Planning Method

7.4.1 Description of the Proposed Method

Figure 7.4 shows the concept of the proposed planning method. Under the proposed method, the production decisions should be reviewed more often because of the dynamics of daily demand, which cannot be forecast with

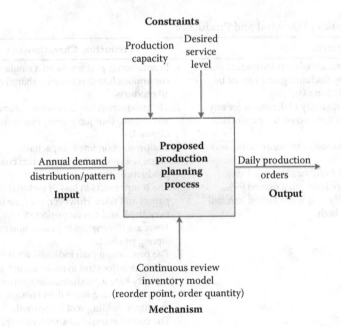

FIGURE 7.4
Overview of the proposed production planning process.

high accuracy. In addition, the continuous inventory review is applied to establish decision guidelines. The inventory policy is calculated with annual demand distribution.

Figure 7.5 shows details of the proposed planning method, which comprises two major components: establishing inventory policy and determining production orders.

The inventory policy is determined by using a simple reorder point order quantity model:

$$\text{Reorder point:} \quad ROP = Z^* \sigma + \mu.$$

$$\text{Order quantity:} \quad Q = \sqrt{\frac{2\lambda(K + \hat{p}^* n(s))}{h}}$$

where

Z = safety stock factor
σ = standard deviation of demand during lead time
μ = mean of demand during lead time
λ = mean of annual demand
K = setup cost
$\hat{p}$ = penalty cost
$n(s)$ = expected shortage = $\sigma * L(Z)$ where $L(Z)$ is the standard loss function
h = inventory holding cost

The inventory policy should be reviewed at least once a year or whenever the demand distribution is changed. It should be noted the production

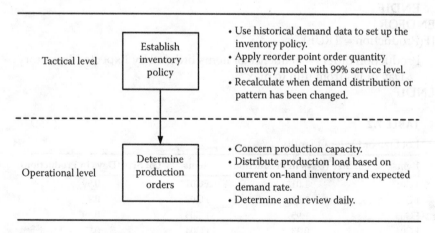

FIGURE 7.5
Components of the proposed planning method.

capacity is not applicable at the tactical level. Table 7.2 shows the order quantity or production lot size in terms of days in production compared to its capacity.

At the operational level, production orders are reviewed and determined every day. At this point, we are facing a multi-item situation on a single production line. Thus the production capacity is considered.

In determining daily production orders, we decide whether or not we should pull expected future production to process beforehand to avoid capacity conflict if we wait until the production orders are triggered by the reorder point (ROP). The pseudo code of our algorithm is shown below.

FOR all *i* in set of FGs

      Expected Inventory Day of FG *i* = Inventory FG *i*/Daily demand *i*

ENDFOR

Production = FALSE

FOR all *i* in set of FGs

  Sum of days in production = 0

  FOR all *j* in set of FGs

    IF (Expected Inventory Day of FG *i* >= Expected Inventory Day of FG *j*)

      Sum of days in production += Production Time *j*

    ENDIF

  ENDFOR

  Capacity of FG *i* = Expected Inventory Day of FG *i* – Sum of days in production

  IF (Capacity of FG *i* < 1)

    Production = TRUE

  ENDIF

ENDFOR

IF(Production = TRUE)

  Production orders = Sequence of items ordered by Expected Inventory Day of FG from min to max

ENDIF

TABLE 7.2

Lot Size or Order Quantity

Items	Lot Size (Days in Production)	Items	Lot Size (Days in Production)
FG01	1.01	FG01	0.99
FG02	1.94	FG02	0.43
FG03	1.95	FG03	0.49
FG04	0.33	FG04	0.77
FG05	0.73	FG05	0.83

TABLE 7.3

Details of Parameters

Type	Parameters	Details
Cost	1. Setup cost	Estimated by the material loss when starting new product production
	2. Inventory holding cost	Estimated by company cost of capital and warehouse cost
	3. Penalty cost	Given by customers
Demand	1. Demand distribution	Statically estimated from actual demand data from the year 2010
Lead time	1. Customer demand	Given by customers
	2. Raw material procurement	Estimated by suppliers and may be varied by types of raw materials

In terms of raw material planning, the base stock model is used to guarantee that there will always be raw materials available whenever they are needed in production. The base stock level is determined by the finished goods production lot size and the expected ratio of production cycle time to procurement lead time. If the ratio is greater than 1, then the raw material base stock value should be enough for one finished goods production lot size. Otherwise, it means that the procurement lead time is greater than the average cycle time; the base stock value then should be greater than 1.

7.4.2 Input Parameters

Input parameters include cost, demand, and lead time parameters. All parameters are estimated from the actual data from the year 2010. The details of parameter estimates are presented in Table 7.3.

7.5 Evaluation and Results

The proposed method is evaluated with a set of actual 3-month demand data from April to June 2011. The initial inventory level is also given with the actual amount. We tested our inventory policy and production planning algorithm in Microsoft Excel®, and then compared results with current planning method. In Figure 7.6, the inventory movement or behavior for each product is illustrated. It shows how each planning method is performed: replenishment decisions, inventory level, and shortages. The raw material performance is shown in Figure 7.7.

The planning performance indicators are reported in Tables 7.4 and 7.5. The indicators include an average inventory level, the number of production line setups for product changed, fill rate, and savings on holding inventory.

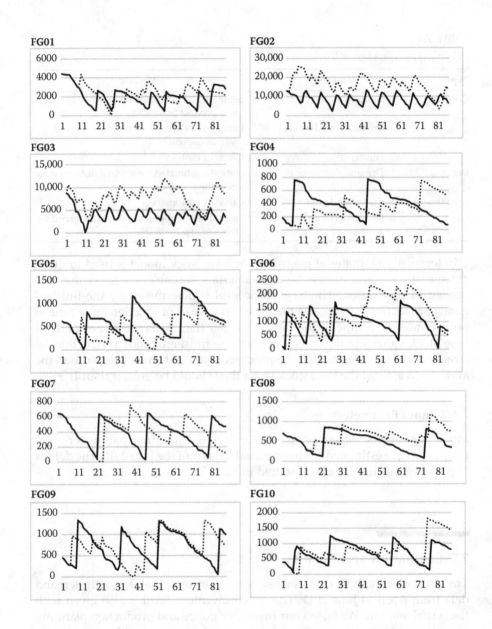

FIGURE 7.6
Computational results for production planning.

   The test results show that with the proposed method, all items can successfully achieve the required service level at 99%, whereas only 3 out of 10
items have achieved this level with the current planning method. In terms
of holding inventory, 8 of 10 items result in a lower level of inventory holding without sacrificing the service level. For best-selling items such as FG02

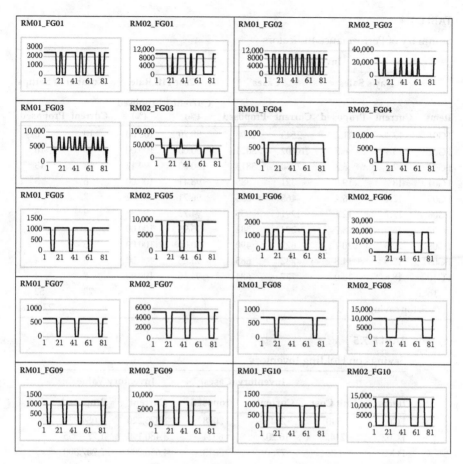

FIGURE 7.7
Computational results for raw materials.

and FG03, the inventory can be reduced as much as almost 50%. For FG04 and FG07, the proposed method results in an inventory increase as a result of larger lot size setup. However, the total setups have been decreased from 10 setups to 6 setups. For almost all items, there is a decrease in the numbers of setups due to fixed lot size policy. Unlike the proposed method, the current method does not have any concrete decision rule to determine lot size. It can be seen in Figure 7.6 that the current production lot sizes are randomly determined. Some lot sizes can be very large, whereas others can be very small. It shows that the current method does not consider economy of scales in a production setup. Furthermore, some items have a decrease in both inventory level and setups. In sum, we can conclude that the proposed method is successful in planning production of the case study's finishing process. It can reduce the overall inventory holding value and production line setups, and the required service level is still achieved.

TABLE 7.4

Summary of Performance Indicators

	Inventory				Case Fill Rate		Number of Setups	
	Day's Sales		Cases		Current (%)	Proposed (%)		
Items	Current	Proposed	Current	Proposed			Current	Proposed
FG01	13.4	11.2	2543	2008	98.70	100.00	10	6
FG02	13.4	6.7	16,202	8024	100.00	100.00	13	10
FG03	12.4	6.8	7760	3813	100.00	99.42	13	13
FG04	18.7	19.8	321	388	98.41	100.00	6	3
FG05	12.1	16.4	465	634	96.71	99.02	7	3
FG06	17.7	12.1	1291	955	94.72	99.55	9	5
FG07	16.3	18.7	394	382	95.95	100.00	4	3
FG08	29.5	20.1	627	490	100.00	100.00	5	2
FG09	15.3	14.8	732	685	95.48	100.00	8	4
FG10	18.6	13.0	777	684	94.16	100.00	8	4
Total							73	53
% Setup savings								27.40%

TABLE 7.5

Savings on Holding Inventory

		Inventory (Cases)		Inventory Value	
Items	Unit Cost	Current	Proposed	Current	Proposed
FG01	540	2543	2008	1,373,220	1,084,320
FG02	214.8	16,202	8024	3,480,190	1,723,555
FG03	600	7760	3813	4,656,000	2,287,800
FG04	705	321	388	226,305	273,540
FG05	567	465	634	263,655	359,478
FG06	620	1291	955	800,420	592,100
FG07	850	394	382	334,900	324,700
FG08	780	627	490	489,060	382,200
FG09	599.3	732	685	438,688	410,521
FG10	932.4	777	684	724,475	637,762
Total				12,786,912	8,075,975
% Savings					36.84%

7.6 Conclusions

Production planning for a rolled tissue finishing process of a large manufacturer is studied. The objective is to improve the current planning method
to achieve its customer expectation. The reorder point and fixed order quantity model are applied in the proposed planning method. The computational

results show that the proposed method can perform well with actual demand data. Future research will focus on improving the algorithm for determining production orders to match various types of demand patterns better.

References

Agrawal, S., R. N. Sengupta, and K. Shanker. Impact of information sharing and lead time on bullwhip effect and on-hand inventory. *European Journal of Operation Research*, 192(2): 576–693, 2009.

Balkhi, Z. Optimal stopping and restarting times for multi item production inventory systems with resource constraints *International Journal of Applied Mathematics*, 3: 27–34, 2009.

Boyaci, T. and G. Gallego. Coordinating pricing and inventory replenishment policies for one wholesaler and one or more geographically dispersed retailers. *International Journal of Production Economics*, 77(2): 95–111, 2002.

Bretthauera, K., B. Shettyb, S. Syamc, and R. Vokurkad. Production and inventory management under multiple resource constraints. *Mathematical and Computer Modeling*, 44: 85–95, 2006.

Buffa, E. and J. Miller. *Production-Inventory Systems: Planning and Control*. Homewood, IL: Irwin, 1979.

DeCroix, G. and A. Arreola-Risa. Optimal production and inventory policy for multiple products under resource constraints. *Management Science*, 44: 950–961, 1998.

Gruen, T. and D. Corsten. Desperately seeking shelf availability: An examination of the extent, the cause and the effects to address retail out-of-stocks. *International Journal of Physical Distribution & Logistics Management*, 31: 605–617, 2003.

Hoque, M. and S. K. Goyal. A heuristic solution procedure for an integrated inventory system under controllable lead-time with equal or unequal sized batch shipments between a vendor and a buyer. *International Journal of Production Economics*, 102(2): 217–225, 2006.

Maloney, B. and C. Klein. Constrained multi-item inventory systems: An implicit approach. *Computers and Operations Research*, 20: 639–649, 1993.

Özer, Ö. and W. Wei. Inventory control with limited capacity and advance demand information. *Operations Research*, 52(6): 988–1000, 2004.

Page, E. and R. Paul. Multi-product inventory situations with one restriction. *Operational Research Quarterly*, 27: 815–834, 1976.

Rego, J. and M. Mesquita. Demand forecasting and inventory control: A simulation study on automotive spare parts. *International Journal of Production Economics*, 161: 1–16, 2015.

Sarmah, S. P., D. Acharya, and S. K. Goyal. Buyer vendor coordination models in supply chain management. *European Journal of Operational Research*, 175: 1–15, 2006.

Zavanella, L. and S. Zanoni. A one-vendor multi-buyer integrated production-inventory model: The 'Consignment Stock' case. *International Journal of Production Economics*, 118: 225–232, 2006.

Section IV

Financial Decisions in the Supply Chain

8

A Game of Competitive Investment: Overcapacity and Underlearning

Jian Yang, Yusen Xia, and Junmin Shi

CONTENTS

ABSTRACT We consider the situation in which a number of firms decide their individual capacity investment levels. The total sum of these levels determines the total return, which the firms share in proportion to their contributions. Before their commitments, firms may spend efforts on learning a size indicator of the market. Using this model, we can explain the overcapacity phenomenon that appeared time and again in numerous industries. The competitive learning aspect of the situation sheds light on the chronic neglect of due diligence when companies are supposed to conduct demand-forecast studies but do not do so.

KEY WORDS: *investment game, overcapacity, underlearning.*

8.1 Introduction

8.1.1 Motivation and Outline

In various product markets, especially those involving huge initial investments but comparatively little operational costs, one may observe that overcapacity, that is, the presence of an industry-wide capacity that is more than desirable for the combined welfare of the producers, is a common phenomenon. In the dynamic random-access memory (DRAM) industry, chip makers have been expanding their production capacities over recent years, resulting in continuous price drops that hurt the makers' profit in turn (Kardos et al., 2008). In the liquid-crystal display (LCD) industry, LG Philips announced a $335 million loss in 2006 due to a steep price decline resulting from an industry-wide capacity glut (Burns, 2006). The auto industry is also plagued by excess capacity. It has been estimated that the global overcapacity in this industry is around 20% (Dressler, 2004).

Despite the prevalence and impact of overcapacity, there is but a dearth of research on the cause of this phenomenon. The few works dealing with this issue treat it as the result of established firms making credible threats to deter the entry of newcomers. However, it is hard to see that the self-harming strategy of overcapacity is used by established firms solely for the purpose of battling the remote chance of new entrants, as initial investments needed to enter the aforementioned industries all involve billions of dollars. It is against this backdrop that we propose a capacity-setting game, involving equally established firms. In this setting, overcapacity comes as a natural competitive outcome as firms jostle for market shares.

Key ingredients of our game-theoretic setting are that (1) the total revenue generated by all firms is increasing and concave in the total capacity built up by all firms; and (2) a firm's revenue share is proportional to its capacity share. These features are consistent with industries in which initial investments are comparatively costlier than day-to-day operations. Our setup naturally leads to the fact that a higher industry-wide capacity is needed in the competitive setting more than in the first-best setting, for a firm's marginal return on capacity investment to be matched by its marginal investment cost. This therefore leads to overcapacity.

Much is at stake when a firm commits to a multiyear project of building a multibillion dollar production facility from scratch while future market outlook is still uncertain. Hence, forecasting of future demand is essential to a firm's survival and prosperity in the face of cutthroat competition. But learning under competition is a tricky business. With the ease at which data travel in this Internet age, it is impossible for a firm to keep what it has learned about the market in complete darkness. When multiple firms collect data about the same market, each participating firm may learn more than it would alone. Yet, as far as we know, there is no conclusive result on whether it positively or negatively affects each player when everybody gains more knowledge about market conditions. Therefore, we may contend that effects of information and learning in competitive environments are not yet well understood, even though benefits of better information have been well established in single-firm settings (e.g., Blackwell, 1951; Lehmann, 1988).

From our capacity game, we take one step toward the better understanding of information and learning under competition. Specifically, we add one stage before the capacity-setting stage. In this stage, firms' efforts are focused on learning a size indicator of the common market. To make matters simple, we let random variables representing the size indicator and signals received by firms be bivalued. Learning is reflected by relationships between these random variables and firms' efforts. Our learning framework reflects the externality in learning, so that the trustworthiness of the signal received by a firm is determined by both the learning effort put in by the current firm and efforts put in by other firms. The framework also allows different firms to receive different signals. As the capacity game involving learning is difficult to analyze in its full generality, we let the investment return function take a special form. We then concentrate on the case in which the information collected by every firm is public knowledge. For this case, we examine the underlearning effect, the phenomenon in which firms shirk from their learning responsibilities, hoping that others will do the dirty work for them.

The following is a summary of our main contributions:

1. We establish that the concavity of the investment return function and the proportionality of revenue allocation are primary culprits for overcapacity (cf. Theorem 8.1).

2. In a competitive setting, we introduce notions for both information structure and controlled learning.

3. For the case in which firms cannot hide what they learn, we demonstrate the severity of the underlearning effect—the more numerous the firms, the less they will know about the market in which they all operate (cf. Theorem 8.3).

Point 1 gives a plausible interpretation to the prevalent overcapacity phenomenon. Also, point 2 provides a functional alternative for modeling information and learning in competitive settings, and point 3 offers firms forewarnings about the dire consequences of neglecting due diligence in their in-house market research before plunging into an uncertain market into which others are rushing as well (e.g., the subprime mortgage market from 2002 to 2007). Finally, we want to add that our underlearning results are consistent with firms' demand learning and capacity investment behaviors in several industries. For example, in the electronics industry, without sufficient and effectively learning demand information, firms invested too much in capacity, which led prices for DRAM chips to fall 70% in 2007 (Ihlwan, 2007).

8.1.2 Literature Survey

In a duopolistic setting, Kreps and Scheinkman (1983) showed that a two-stage game involving capacity competition in the first stage and Bertrand-like price competition in the second stage results in a Cournot-like equilibrium. Davidson and Deneckere (1986) pointed out the earlier result's critical dependence on a particular demand rationing rule. Acemoglu et al. (2009) considered a similar two-stage model in which consumers always exhaust a lower priced firm's capacity before moving on to a higher priced firm. They quantified inefficiencies of the game's equilibria and demonstrated that great differences exist between different equilibria. Anupindi and Jiang (2008) studied the role played by flexibility in a duopolistic game involving capacity decisions. The industrial organization literature has shown that excess capacity can be exploited by an established firm as a credible threat to deter entry; see, for example, Dixit (1980) and Bulow et al. (1985).

The competitive-newsboy framework of Lippman and McCardle (1997) can certainly be used in a competitive-capacity study. However, this framework is more suitable for the situation in which the set of firms under scrutiny constitutes only a small portion of the entire industry, to the effect that the total capacity built by these firms does not have any sway over the sales price of the concerned product. Incidentally, as examined in Cachon (2003, Section 6.5.1), this framework also leads to an overcapacity phenomenon, resulting primarily from the fact that each firm ignores the demand-reducing effect on other firms when it builds excess capacity. The same reason is behind the overcapacity effect identified by Mahajan and van Ryzin (2001) in their dynamic consumer choice framework. Our study of overcapacity across entire industries necessitates a different setup.

Elastic demand was indeed considered in Deneckere et al. (1997) and Cachon (2003, Section 6.5.2). However, their models allow for an unlimited supply of nonatomic firms. New firms will come into competition as long as the market price has not been driven to zero. Our setting is oligopolistic with a fixed number of firms. We shall demonstrate overcapacity at the individual-firm level.

To leave room for the later learning-stage addition, we do not explicitly model a pricing-and-rationing stage after the capacity-setting stage. Nevertheless, our setup takes into account the industry-wide capacity's dampening effect on firms' pricing powers. As mentioned, our fundamental assumptions on the capacity-setting stage are the concave total revenue function and proportional revenue allocation. One explanation for these assumptions is as follows: When production is relatively cheap, firms tend to fully utilize their capacities. This way, the total industry-wide capacity, together with the innate demand-price curve of the market, will determine the market-clearing price, and hence the total industry-wide revenue. Because all firms face the same price, each firm's revenue share is its share of the total capacity.

Under this setup, a firm has the incentive to keep on expanding its capacity until the marginal return on its own buildup can no longer offset the required investment. With revenue being shared proportionally, a firm's marginal return will be more diluted when there are more competitors in the market. On the other hand, it takes a higher capacity for a diluted marginal return to match its undiluted counterpart. This is the main reason behind the overcapacity phenomenon.

The literature on learning in monopolistic settings is extensive. For instance, Burnétas and Gilbert (2001) showed that demand learning can help reduce procurement costs; under various circumstances, Lariviere and Porteus (1999), Bensoussan et al. (2007), and Chen and Plambeck (2008) demonstrated that Bayesian learning can help firms cope with unobserved lost sales. A few works in economics dealt with "learning by doing" in competitive settings. Rob (1991) treated a multiperiod rational expectations model in which firms base their decisions of entering and exiting a market on past information generated by incumbent firms' actions. Aghion et al. (1993) studied a multiperiod pricing game involving firms that produce differentiated products, whereby all firms receive the same information on the market demand which is influenced by past actions of all firms. These works emphasized the "public good" aspect of information and their equilibria often exhibit the "free-riding" phenomenon.

Our setup for information and learning can demonstrate the aforementioned informational externality features as well. Furthermore, it has improved over existing frameworks in the aspects that learning has been made explicit and separately controllable. That is, "learning" is no longer entangled with "doing." This way, investments in market studies and experiments can be modeled directly, and trade-offs between information-acquisition costs and gains due to better information can be more clearly analyzed; in addition, our framework can be more readily transplanted to different settings. Shin and Tunca (2009) showed that overlearning can occur when retailers ordering from one single supplier are themselves engaged in Cournot competition. However, they assumed that there is no externality in learning, and hence every retailer is singularly responsible for its own learning. When retailers' learning efforts (but not the signals they acquired) become public, the same authors demonstrated that the overlearning effect will be amplified further. One of our main results is almost complementary. It says that, when all firms receive the same signal produced by their collective efforts, firms will tend to shirk from their forecasting responsibilities. Also, we assume that the efforts put into learning are public knowledge throughout.

The rest of the chapter is organized as follows. In Section 8.2, we set up the capacity investment game and provide basic analyses; in Section 8.3, we introduce notions of information structure and controlled learning; in Section 8.4, we analyze the underlearning effect; in Section 8.5, we shed light on a potential extension to the case where different firms may acquire different signals; and finally, we conclude the chapter in Section 8.6.

8.2 The Overcapacity Effect

8.2.1 Setup

Our investment game involves n identical firms. These firms compete in utilizing costly capital to build capacities with hopes of generating future returns. Each firm's cost for capital follows a function $c: R^+ \to R^+$, where R^+ stands for $[0 + \infty)$. The return to an individual firm is not solely determined by its own investment level. Rather, the total return to all firms is governed by a function $r: R^+ \to R^+$ of these firms' total investment level. The return to each firm i is proportional to its investment level $x_i \in R^+$. Therefore, when the profile of other firms' investment levels is $x_i = (x_j | j \neq i)$, firm i will receive a profit $f(x_i, x_{-i})$, where

$$f(x_i, x_{-i}) = \frac{x_i}{x_i + \sum_{j \neq i} x_j} \cdot r\left(x_i + \sum_{j \neq i} x_j\right) - c(x_i). \tag{8.1}$$

We now give one potential explanation to the aforementioned setup. Suppose the demand function of the concerned product is given by $d = D(p)$, whose inverse is $p = P(d)$. Also, there is one production run after the capacity buildup. Finally, suppose that, relative to $c'(0)$, the unit production cost c_p is negligible. This way, firms will tend to produce at full capacity. The total supply on the market will be the total capacity level $\sum_{i=1}^{n} x_i$, which will lead to a market-clearing price $P\left(\sum_{i=1}^{n} x_i\right)$. The total revenue made by all firms will therefore be

$$r\left(\sum_{i=1}^{n} x_i\right) = \left(\sum_{i=1}^{n} x_i\right) \cdot P\left(\sum_{i=1}^{n} x_i\right), \tag{8.2}$$

while firm i's share of the total revenue will be $\left(x_i / \left(x_i + \sum_{j \neq i} x_j\right)\right) \cdot r\left(x_i + \sum_{j \neq i} x_j\right)$. When there are an infinite number of production stages after the settlement of capacity levels, where the unit production cost at every stage is c_p and the per-stage discount factor is δ, we will need $c_p/(1 \delta) = c'(0)$ for the preceding explanation to work.

Owing to the apparent symmetry in Equation 8.1, we may define function g, so that

$$g(x,y) = \frac{x}{x+y} \cdot r(x+y) - c(x). \tag{8.3}$$

Note that

$$f(x_i, x_{-i}) = g\left(x_i, \sum_{j \neq i} x_j\right). \tag{8.4}$$

We suppose that the capital cost function c is smooth. In addition, we assume the following:

(c0) $c(0) = 0$, as is expected

(c1) $c'(0^+) \geq 1$, which reflects the lost opportunity of invested capital

(c2) $c''(x) \geq 0$ for any $x \in (0, +\infty)$, so that the marginal cost of capital is increasing

We further suppose that the return function r is smooth. In addition, we assume the following:

(r0) $r(0) = 0$, as is expected

(r1) $r'(x) > 0$ for any $x \in (0, +\infty)$, so that more investment leads to a higher return

(r2) $r''(x) < 0$ for any $x \in (0, +\infty)$, so that the marginal rate of return to investment decreases with the investment level

(r3) $\lim_{x \to +\infty} r'(x) = 0$, so that return to investment will diminish to zero when capital is injected in indefinitely

Between functions c and r, we assume that

(cr) $r'(0^+) > c'(0^+)$, so that investment will be lucrative when the industry-wide capacity is low enough.

One immediate consequence of (r0) and (r2) is that $r'(x) < (r(x) - r(0))/(x - 0) = r(x)/x$. We put this in the following:

(r02) $r'(x) < r(x)/x$ for any $x \in (0, +\infty)$.

A further consequence of the preceding is that $(r(x)/x)' = (r'(x) - r(x)/x)/x < 0$. Therefore, we have

(r02b) $r(x)/x$ is decreasing in x.

8.2.2 Competitive Analysis

From Equation 8.3, we have

$$\frac{\partial g(x,y)}{\partial x} = \frac{x}{x+y}r'(x+y) + \frac{y}{(x+y)^2}r(x+y) - c'(x), \tag{8.5}$$

and

$$\frac{\partial^2 g(x,y)}{\partial x^2} = \frac{x}{x+y}r''(x+y) + \frac{2y}{(x+y)^2}\left[r'(x+y) - \frac{r(x+y)}{x+y}\right] - c''(x). \tag{8.6}$$

From (r2), (r02), (c2), and Equation 8.6, we know that $\partial^2 g(x,y)/\partial x^2 < 0$, and hence $g(x, y)$ is strictly concave in x.

We set out to see if a pure symmetric equilibrium $(x_i = x_n^* \mid i = 1, 2, ..., n)$ exists. Here, the subscript "n" in "x_n^*" signifies the presence of n firms. From the strict concavity of $g(x, y)$ in x and Equation 8.5, we see that x_n^* can be found by solving the following:

$$h_n\left(x_n^*\right) = \frac{\partial g(x,y)}{\partial x}\bigg|_{x=x_n^*,\, y=(n-1)x_n^*} = \frac{1}{n}\cdot r'(nx_n^*) + \frac{n-1}{n}\cdot\frac{r\left(nx_n^*\right)}{nx_n^*} - c'\left(x_n^*\right) = 0, \tag{8.7}$$

whenever the equality is achievable. The desired x_n^* is not only in existence, but also unique.

Proposition 8.1

There is a unique pure symmetric equilibrium investment level x_n^*.   ∎

We have relegated all proofs to appendices that follow the main text. For instance, the proof of Proposition 8.1 appears in Appendix A. Now we study the total investment level $z_n^* = nx_n^*$. To this end, define function i_n so that $i_n(z) = h_n(z/n)$. By Equation 8.7, we have

$$i_n(z) = (1/n)\cdot r'(z) + ((n-1)/n)\cdot r(z)/z - c'(z/n)$$
$$= r(z)/z + (1/n)\cdot[r'(z) - r(z)/z] - c'(z/n). \tag{8.8}$$

Being a rescaled version of h_n, the function i_n is positive at 0^+, decreasing in z, and negative at large z values. Just because x_n^* is the unique root of h_n, we know that z_n^* is the unique root of i_n. More importantly, we know that the marketwise investment level z_n^* increases with the number of participants n.

Proposition 8.2

The equilibrium total investment level z_n^* is increasing in n. ∎

8.2.3 Comparison with the First-Best Solution

For the total payoff of the n firms to be maximized, a social planner will solve the following problem:

$$\max \quad r(z) - n \cdot c(z/n)$$
$$\text{s.t.} \quad z \in R^+. \tag{8.9}$$

Therefore, the first-best total investment level z_{1n}^* will be a solution to

$$i_{1n}(z) = r'(z) - c'\left(\frac{z}{n}\right) = 0. \tag{8.10}$$

Here, the subscript "1" signifies optimality under one decision maker, and the subscript "n" still signifies the presence of n firms. By (r2) and (c2), we know that $i_{1n}(z)$ is strictly decreasing in z; by (cr), we know that $i_{1n}(0^+) > 0$; and, by (r3), (c1), and (c2), we know that $i_{1n}(z) < 0$ will occur when z is large enough. Therefore, z_{1n}^* is in existence and unique. Moreover, we can predict the trend for z_{1n}^* when the number of participants n changes.

Proposition 8.3

The first-best total investment level z_{1n}^* is increasing in n, while the first-best individual investment level $x_{1n}^* = z_{1n}^*/n$ is decreasing in n. ∎

We can show the important result that competition brings in overcapacity.

Theorem 8.1

Compared to the first-best total investment level z_{1n}^*, the equilibrium level z_n^* is greater. ∎

Our earlier assumptions and the above result essentially reflect that over-capacity is a consequence of the concavity of the investment return function and the proportionality of revenue allocation. In general, the difference function $(i_n - i_{1n})$ as defined in Equation 8.60 indicates the degree of overcapacity. We may see that it increases with n, and saturates at the function $(r(z)/z - r'(z))$ as n tends to $+\infty$.

As a specific case, suppose $r(x) = x^\gamma$ for some $\gamma \in (0,1)$ and $c(x) = x$. Then, from Equations 8.8 and 8.10, we may find that $z_n^* = ((n-1)/n + \gamma/n)^{1/(1-\gamma)}$ and $z_{1n}^* = \gamma^{1/(1-\gamma)}$. Hence, a more direct measure of overcapacity, $z_n^*/z_{1n}^* = ((n-1)/(n\gamma) + 1/n)^{1/(1-\gamma)}$, grows with n fairly quickly. When $\gamma = 1/2$, it follows that $z_n^*/z_{1n}^* = (2n-1)^2/n^2$, which converges to 4 as $n \to +\infty$.

8.3 Information Structure and Learning

We introduce a framework that allows uncertainty in the size of the market faced by all firms. Through individual efforts, firms may exert control on the precision levels of the common signal received by all of them.

8.3.1 A Learning Framework

Instead of the previous r, let now the return function R_ω be parameterized by some $\omega \in \{L, H\}$, where $H > L > 0$. Before obtaining any information, all firms believe that ω is the realization of the random variable Ω, which satisfies

$$P[\Omega = L] = P[\Omega = H] = \frac{1}{2}. \tag{8.11}$$

To describe the information structure to be used, we introduce random variable Θ, which serves as a common signal that reflects the commonality of firms' knowledge about Ω. More specifically, we let Θ be a bivalued random variable ranging in $\{L, H\}$. There is a constant $a \in [0, 1]$ such that

$$\begin{cases} P[\Theta = L \mid \Omega = L] = (1+a)/2 = P[\Theta = H \mid \Omega = H], \\ P[\Theta = L \mid \Omega = H] = (1-a)/2 = P[\Theta = H \mid \Omega = L]. \end{cases} \tag{8.12}$$

From the preceding, we may derive that

$$P[\Omega = L] = P[\Omega = H] = P[\Theta = L] = P[\Theta = H] = \frac{1}{2}. \tag{8.13}$$

Our framework can be extended beyond the current bi-valued case. We opt for the current case for the ease of later derivations involving controlled information acquisition.

Now, we suppose that the information structure itself is subject to firms' controls. That is, we allow the random variable Θ to be dependent on firms' efforts. There are stages 0 and 1. In stage 0, firms may invest to learn the market; in stage 1, they may participate in the investment game described in Section 8.2. Before stage 0, all firms believe that the return function follows $R_\Omega(x)$. Given stage 0 firm-effort vector $x_0 = (x_{0i} | i = 1,2,\ldots,n)$, we suppose that Θ is in the form of $\tilde{\Theta}(x_0)$. Thus, $\tilde{\Theta}(x_0)$ reflects the learning effect. The true value of Ω will be revealed only after stage 1.

We suppose that the a used in Equation 8.12 is replaced by some function $\tilde{a}(x_0)$. With effort-dependent substitutions, we can achieve counterparts of the earlier Equation 8.12. For the function $\tilde{a}(\cdot)$, we suppose that there is a positive constant α, so that

$$\tilde{a}(x_0) = \frac{\alpha \sum_{j=1}^{n} x_{0j}}{1+\alpha \sum_{j=1}^{n} x_{0j}}. \tag{8.14}$$

This function form reflects that more can be learned through the exertion of greater efforts, and that the marginal return in learning decreases with effort levels. We have simplified the matter by letting $\tilde{a}(x_0)$ depend on $\sum_{j=1}^{n} x_{0j}$ only. Because of this, we later write $\tilde{a}\left(\sum_{j=1}^{n} x_{0j}\right)$ in the place of $\tilde{a}(x_{01},\ldots,x_{0n})$.

Also in the preceding, α indicates the effectiveness of learning. At the extreme of $\alpha = 0$, the intermediate signal $\tilde{\Theta}(x_0)$ will be useless noise regardless of the amount of effort spent; at the other extreme of $\alpha = +\infty$, $\tilde{\Theta}(x_0)$ will be Ω itself under the convention that $+\infty \cdot 0 = +\infty$.

8.3.2 Particulars of the Investment Game

Let us specify the particular form of the parameterized return function $R_\omega(\cdot)$:

$$R_\omega(x) = \omega \cdot r\left(\frac{x}{\omega}\right), \quad \forall x \in R^+. \tag{8.15}$$

Note that

$$\frac{dR_\omega(y)}{dy}\Big|_{y=\omega x} = \omega \cdot \frac{1}{\omega} \cdot r'(z)\Big|_{z=y/\omega=\omega x/\omega=x} = r'(x). \tag{8.16}$$

That is, the marginal return of an ωx-level investment under the $R_\omega(\cdot)$-return regime is the same as the marginal return of an x-level investment under the $r(\cdot)$-return regime. Thus, ω in $R_\omega(\cdot)$ can be thought of as a market-size indicator. Now, we make the simplifying assumption that the total cost to firm i is $x_{0i} + x_{1i}$ when it has spent learning effort x_{0i} and capacity investment x_{1i} in the two stages. Now (cr) in Section 8.2 means that $r'(0^+) > 1$.

We may use $x_i = (x_{0i}, x_{1i}(L), x_{1i}(H))$ to describe firm i's strategy. In it, x_{0i} is the firm's stage 0 learning effort, while $x_{1i}(\theta)$ is its stage 1 investment level when it has learned θ as the realization of $\tilde{\Theta}(x_0)$. Let us use $f(x_i, x_{-i})$ to describe the average payoff to firm i, when it adopts policy $x_i = (x_{0i}, x_{1i}(L), x_{1i}(H))$ while others have adopted policy profile $x_i = ((x_{0j}, x_{1j}(L), x_{1j}(H)) | j \neq i$. We have

$$f(x_i, x_{-i}) = \sum_{\theta \in \{L,H\}} (1/2) \cdot \sum_{\omega \in \{L,H\}} q(\omega | \theta; x_{0i}, x_{0,-i}) \times [(x_{1i}(\theta)/(x_{1i}(\theta)$$

$$+ \sum_{j \neq i} x_{1j}(\theta))) \cdot \omega \cdot r((x_{1i}(\theta) + \sum_{j \neq i} x_{1j}(\theta))/\omega) - x_{0i} - x_{1i}(\theta)]\}, \tag{8.17}$$

where, according to the effort-dependent version of Equation 8.12,

$$q(\omega | \theta; x_{0i}, x_{0,-i}) = \begin{cases} \left(1 + 2\alpha x_{0i} + 2\alpha \sum_{j \neq i} x_{0j}\right) \Big/ \left(2\left(1 + \alpha x_{0i} + \alpha \sum_{j \neq i} x_{0j}\right)\right), & \text{when } \theta = \omega, \\[3mm] 1 \Big/ \left(2\left(1 + \alpha x_{0i} + \alpha \sum_{j \neq i} x_{0j}\right)\right), & \text{when } \theta \neq \omega. \end{cases} \tag{8.18}$$

Inside Equation 8.17, the $1/2$ is the chance for the common observation Θ to be either L or H, and $q(\omega | \theta; x_{0i}, x_{0,-i})$ is the conditional probability $P[\Omega = \omega | \Theta = \theta]$ under the same learning-effort vector. Note that firm i's decision dependent is only on θ, whereas its payoff is dependent on the actual Ω-realization ω. As efforts are measured in costs, the cost term $(x_{0i} + x_{1i}(\theta))$, involving unit coefficients, in no way indicates that information acquisition and capacity expansion costs are comparable.

8.4 The Underlearning Effect

By Equation 8.18, we can condense other firms' action profile x_{-i} into $y = (y_0, y_1(L), y_1(H))$ with $y = \sum_{j \neq i} x_j$, meaning, component-wise, that $y_0 = \sum_{j \neq i} x_{0j}$, $y_1(L) = \sum_{j \neq i} x_{1j}(L)$, and $y_1(H) = \sum_{j \neq i} x_{1j}(H)$.

8.4.1 Stage 1 Competitive Analysis

Define function $g(x, y)$, so that $g\left(x_i, \sum_{j \neq i} x_j\right) = f(x_i, x_{-i})$ as defined in Equation 8.17. For ease of presentation, we shall use x_0, x_L, x_H to represent x_0, $x_1(L)$, $x_1(H)$, respectively, and do the same for the y's. Hence, x_0 stands for the current firm's stage 0 learning effort and y_0 stands for other firms' total stage 0 learning effort; for $\theta = L, H$, x_θ stands for the current firm's stage 1 investment level and y_θ stands for other firms' total stage 1 investment level.

Now Equation 8.17 can be simplified into

$$
\begin{aligned}
g(x,y) = g(x_0,x_L,x_H,y_0,y_L,y_H) = &-x_0 - x_L/2 - x_H/2 \\
&+ [(1+2\alpha x_0 + 2\alpha y_0)/(4+4\alpha x_0 + 4\alpha y_0)] \\
&\times [(x_L/(x_L+y_L)) \cdot L \cdot r((x_L+y_L)/L) + (x_H/(x_H+y_H)) \cdot H \cdot r((x_H+y_H)/H)] \\
&+ [1/(4+4\alpha x_0 + 4\alpha y_0)] \\
&\times [(x_L/(x_L+y_L)) \cdot H \cdot r((x_L+y_L)/H) + (x_H/(x_H+y_H)) \cdot L \cdot r((x_H+y_H)/L)].
\end{aligned}
$$

(8.19)

From this, we can derive that

$$
\frac{\partial g(x,y)}{\partial x_L} = g_L'(x_0 + y_0, x_L, y_L),
$$

(8.20)

where

$$
\begin{aligned}
g_L'(z_0, x_L, y_L) = &-1/2 + [(1+2\alpha z_0)/(4+4\alpha z_0)] \\
&\times [(x_L/(x_L+y_L)) \cdot r'((x_L+y_L)/L) + (y_L/(x_L+y_L)^2) \cdot L \cdot r((x_L+y_L)/L)] \\
&+ [1/(4+4\alpha z_0)] \times [(x_L/(x_L+y_L)) \cdot r'((x_L+y_L)/H) \\
&+ (y_L/(x_L+y_L)^2) \cdot H \cdot r((x_L+y_L)/H)].
\end{aligned}
$$

(8.21)

Symmetrically, we can find the expression for $\partial g(x,y)/\partial x_H = g_H'(x_0 + y_0, x_H, y_H)$. These lead to the following important property.

Lemma 8.1

We have $\partial g_L'(z_0, x_L, y_L)/\partial x_L < 0$ *and* $\partial g_H'(z_0, x_H, y_H)/\partial x_H < 0$. ∎

We seek a pure symmetric equilibrium $(x_i = x_n^* \mid i = 1, 2, ..., n)$, where x_n^* is made up of three components: x_{n0}^*, x_{nL}^*, and x_{nH}^*. For the time being, we concentrate on firms' stage-1 subgame perfect equilibrium actions when all firms' stage 0 actions are known. Let $\tilde{x}_{nL}^*(z_0)$ be each firm's pure symmetric equilibrium stage 1 action when it is known that the total learning effort in stage 0 is z_0 and the firm itself has observed the L signal. We may similarly define $\tilde{x}_{nH}^*(z_0)$. Once the stage 0 equilibrium x_{n0}^* has been determined, we can let $x_{nL}^* = \tilde{x}_{nL}^*(nx_{n0}^*)$ and $x_{nH}^* = \tilde{x}_{nH}^*(nx_{n0}^*)$.

By Lemma 8.1, we know that, under a given total stage 0 learning effort z_0, the pair $(\tilde{x}_{nL}^*(z_0), \tilde{x}_{nH}^*(z_0))$ can be found by solving for

$$g_L'(z_0, \tilde{x}_{nL}^*(z_0), (n-1)\tilde{x}_{nL}^*(z_0)) = 0, \quad g_H'(z_0, \tilde{x}_{nH}^*(z_0), (n-1)\tilde{x}_{nH}^*(z_0)) = 0, \quad (8.22)$$

whenever the aforementioned equalities are achievable. According to Equation 8.21, the above requirement on $\tilde{x}_{nL}^*(z_0)$ is that it be a root for function $h_{nL}(z_0, \cdot)$ defined by

$$h_{nL}(z_0, x_L) = -\frac{1}{2} + \frac{1 + 2\alpha z_0}{4 + 4\alpha z_0} \cdot j_n\left(\frac{nx_L}{L}\right) + \frac{1}{4 + 4\alpha z_0} \cdot j_n\left(\frac{nx_L}{H}\right), \quad (8.23)$$

where

$$j_n(x) = \frac{1}{n} \cdot r'(x) + \frac{n-1}{n} \cdot \frac{r(x)}{x}. \quad (8.24)$$

By (r2) and (r02b), we know that $j_n(x)$ is strictly decreasing in x. Symmetrically, we may resort to function $h_{nH}(z_0, \cdot)$ to find $\tilde{x}_{nH}^*(z_0)$, where

$$h_{nH}(z_0, x_H) = -\frac{1}{2} + \frac{1 + 2\alpha z_0}{4 + 4\alpha z_0} \cdot j_n\left(\frac{nx_H}{H}\right) + \frac{1}{4 + 4\alpha z_0} \cdot j_n\left(\frac{nx_H}{L}\right). \quad (8.25)$$

We can establish the existence and uniqueness of a stage 1 subgame perfect equilibrium.

Proposition 8.4

On $(0, +\infty)$, both $h_{nL}(z_0, \cdot)$ and $h_{nH}(z_0, \cdot)$ are strictly decreasing functions starting with strictly positive values and ending with strictly negative values. Consequently, when firms contribute a total learning effort z_0, there will be a unique low-signal investment level $\tilde{x}_{nL}^*(z_0)$ and a unique high-signal investment level $\tilde{x}_{nH}^*(z_0)$ all will be willing to adopt. ∎

From the following, we see that when better informed, firms will place more trust in the common signal they receive, and more boldly place differentiated bets on the market.

Proposition 8.5

First, we have $\tilde{x}_{nL}^*(0) = \tilde{x}_{nH}^*(0)$. Then, as z_0 increases, $\tilde{x}_{nL}^*(z_0)$ will decrease, while $\tilde{x}_{nH}^*(z_0)$ will increase. ∎

Using the same logic as previously, we can show that increases in learning effectiveness α will result in the widening of the gap between $\tilde{x}_{nL}^*(z_0)$ and $\tilde{x}_{nH}^*(z_0)$.

Proposition 8.6

First, we have $\tilde{x}_{nL}^*(z_0) = \tilde{x}_{nH}^*(z_0)$ when $\alpha = 0$. Then, as α increases, $\tilde{x}_{nL}^*(z_0)$ will decrease, while $\tilde{x}_{nH}^*(z_0)$ will increase. ∎

8.4.2 Stage 0 Competitive Analysis

We now come back to stage 0 to study its equilibrium decision. Define function $\tilde{g}_n(x_0, y_0)$, so that

$$\tilde{g}_n(x_0, y_0) = g(x_0, \tilde{x}_{nL}^*(x_0 + y_0), \tilde{x}_{nH}^*(x_0 + y_0), y_0, (n-1)\tilde{x}_{nL}^*(x_0 + y_0),$$
$$(n-1)\tilde{x}_{nH}^*(x_0 + y_0)), \tag{8.26}$$

where g is given in Equation 8.19. The newly defined function is the payoff to a firm which spends an x_0 effort in stage 0, when the other $n-1$ firms spend a total of y_0 effort in this stage and all firms adopt their subgame perfect equilibrium responses in stage 1. By Equation 8.19, we have

$$\tilde{g}_n(x_0, y_0) = -x_0 - \tilde{x}_{nL}^*(x_0 + y_0)/2 - \tilde{x}_{nH}^*(x_0 + y_0)/2$$

$$+ [(1 + 2\alpha x_0 + 2\alpha y_0)/(4 + 4\alpha x_0 + 4\alpha y_0)] \times [(L/n) \cdot r(n\tilde{x}_{nL}^*(x_0 + y_0)/L)$$

$$+ (H/n) \cdot r(n\tilde{x}_{nH}^*(x_0 + y_0)/H)] + [1/(4 + 4\alpha x_0 + 4\alpha y_0)]$$

$$\times [(H/n) \cdot r(n\tilde{x}_{nL}^*(x_0 + y_0)/H) + (L/n) \cdot r(n\tilde{x}_{nH}^*(x_0 + y_0)/L)]. \tag{8.27}$$

Because the above $\tilde{g}_n$ is difficult to analyze, we shall start to make the simplifying assumption:

$$r(x) = \sqrt{x}. \tag{8.28}$$

Now, from Equation 8.24, we will have

$$j_n(x) = \frac{2n-1}{2n\sqrt{x}}. \tag{8.29}$$

Hence, from Equations 8.23 and 8.25, we have

$$\begin{cases} h_{nL}(z_0, x_L) = -1/2 + ((2n-1)/(8n+8n\alpha z_0)) \\ \qquad \times [(1+2\alpha z_0) \cdot \sqrt{L/(nx_L)} + \sqrt{H/(nx_L)}], \\ h_{nH}(z_0, x_H) = -1/2 + ((2n-1)/(8n+8n\alpha z_0)) \\ \qquad \times [(1+2\alpha z_0) \cdot \sqrt{H/(nx_H)} + \sqrt{L/(nx_H)}], \end{cases} \tag{8.30}$$

from which we get

$$\begin{cases} \tilde{x}_{nL}^*(z_0) = (2n-1)^2 \times (\sqrt{L} + \sqrt{H} + 2\sqrt{L} \cdot \alpha z_0)^2 / (16n^3 \cdot (1+\alpha z_0)^2), \\ \tilde{x}_{nH}^*(z_0) = (2n-1)^2 \times (\sqrt{H} + \sqrt{L} + 2\sqrt{H} \cdot \alpha z_0)^2 / (16n^3 \cdot (1+\alpha z_0)^2). \end{cases} \tag{8.31}$$

Plugging Equations 8.28 and 8.31 into 8.27, we obtain a closed-form expression for $\tilde{g}_n$:

$$\tilde{g}_n(x_0, y_0) = -x_0 + [(2n-1)/(16n^3 \cdot (1+\alpha x_0 + \alpha y_0)^2)]$$
$$\times \left\{ 2\sqrt{HL} \cdot (1+2\alpha x_0 + 2\alpha y_0) + (L+H) \cdot [1+2\alpha(x_0+y_0) \cdot (1+\alpha x_0 + \alpha y_0)] \right\}. \tag{8.32}$$

After some algebra, it can be found that

$$\frac{\partial \tilde{g}_n(x_0, y_0)}{\partial x_0} = Q(A_n, \alpha(x_0 + y_0)), \tag{8.33}$$

where

$$Q(a, w) = \frac{aw}{(1+w)^3} - 1, \tag{8.34}$$

and

$$A_n = \frac{(2n-1)\alpha\left(\sqrt{H} - \sqrt{L}\right)^2}{8n^3}.$$ (8.35)

At various a levels, the function $Q(a,\cdot)$ possesses useful properties.

Lemma 8.2

When $a \in [0, 27/4)$, we have $Q(a, w) < 0$ for $w \in R^+$; when $a \in [27/4, +\infty)$, the function $Q(a,\cdot)$ has two positive real roots $w^0(a) \leq 1/2$ and $w^*(a) \geq 1/2$, such that the function is below 0 when $w \in [0, w^0(a))$, above 0 when $w \in [w^0(a), w^*(a)]$, and below 0 again when $w \in (w^*(a), +\infty)$. ∎

Suppose parameters α, L, and H are such that the A_n as defined by Equation 8.35 is above 27/4. Then, we may define $B_n = w^*(A_n)$, where $w^*(a)$ is the larger one of the two roots of function $Q(a,\cdot)$ as defined in Lemma 8.2. Also, define C_n so that

$$C_n = n\alpha \cdot \left[\tilde{g}_n\left(\frac{B_n}{n\alpha}, \frac{(n-1)B_n}{n\alpha}\right) - \tilde{g}_n\left(0, \frac{(n-1)B_n}{n\alpha}\right) \right],$$ (8.36)

which, after some algebra while using Equation 8.32 and the fact that $Q(A_n, B_n) = Q(A_n, w^*(A_n) = 0)$, can be found to be the same as

$$C_n = \frac{(2n-2)\cdot B_n^3 + nB_n \cdot (B_n - 1)}{2\cdot(nB_n - B_n + n)^2}.$$ (8.37)

We can express the largest stage 0 equilibrium x_{n0}^* in terms of the aforementioned constants.

Proposition 8.7

x_{n0}^* *is always in existence; in addition, it is true that*

$$x_{n0}^* = \begin{cases} 0, & \text{when } A_n < 27/4, \text{ or } A_n \geq 27/4 \text{ and } C_n < 0, \\ B_n/(n\alpha), & \text{when } A_n \geq 27/4 \text{ and } C_n \geq 0. \end{cases}$$
∎

8.4.3 First-Best Analysis

Suppose all firms adopt the same policy $x = (x_0, x_L, x_H)$. Then, according to Equation 8.19, each of the firms will earn the following:

$$g_1(x_0, x_L, x_H) = g(x_0, x_L, x_H, (n-1)x_0, (n-1)x_L, (n-1)x_H) = -x_0 - x_L/2$$
$$- x_H/2 + [(1 + 2n\alpha x_0)/(4 + 4n\alpha x_0)]$$
$$\times \left[(L/n) \cdot \sqrt{nx_L/L} + (H/n) \cdot \sqrt{nx_H/H} \right]$$
$$+ [1/(4 + 4n\alpha x_0)] \times \left[(H/n) \cdot \sqrt{nx_L/H} + (L/n) \cdot \sqrt{nx_H/L} \right]. \quad (8.38)$$

Hence, we have

$$\frac{\partial g_1(x_0, x_L, x_H)}{\partial x_L} = g'_{1L}(nx_0, x_L), \quad (8.39)$$

where

$$g'_{1L}(z_0, x_L) = -\frac{1}{2} + \frac{(1 + 2\alpha z_0) \cdot \sqrt{L} + \sqrt{H}}{(8 + 8\alpha z_0) \cdot \sqrt{nx_L}}. \quad (8.40)$$

It is easy to check that $g'_{1L}(z_0, \cdot)$ is strictly decreasing in x_L, $\lim_{x_L \to 0^+} g'_{1L}$ $(z_0, x_0) > 0$, and $\lim_{x_L \to +\infty} g'_{1L}(z_0, x_L) < 0$. We can find similar properties for $\partial g_1(x_0, x_L, x_H)/\partial x_H = g'_{1H}(nx_0, x_H)$. Therefore, given total learning effort z_0, the stage 1 first-best decisions $\tilde{x}^*_{1nL}(z_0)$ and $\tilde{x}^*_{1nH}(z_0)$ can be found by solving for $g'_{1L}(z_0, \tilde{x}^*_{1nL}(z_0)) = 0$ and $g'_{1H}(z_0, \tilde{x}^*_{1nH}(z_0)) = 0$, respectively. Hence, in view of Equation 8.40, we have

$$\begin{cases} \tilde{x}^*_{1nL}(z_0) = \left(\sqrt{L} + \sqrt{H} + 2\sqrt{L} \cdot \alpha z_0 \right)^2 / (16n \cdot (1 + \alpha z_0)^2), \\ \tilde{x}^*_{1nH}(z_0) = \left(\sqrt{H} + \sqrt{L} + 2\sqrt{H} \cdot \alpha z_0 \right)^2 / (16n \cdot (1 + \alpha z_0)^2). \end{cases} \quad (8.41)$$

Comparing Equations 8.31 and 8.41, we may see that overcapacity is present under the same level of learning. Also, the ratio of overcapacity approaches 4 from below when the number of firms tends to $+\infty$.

Theorem 8.2

We have

$$\frac{\tilde{x}^*_{nL}(z_0)}{\tilde{x}^*_{1nL}(z_0)} = \frac{\tilde{x}^*_{nH}(z_0)}{\tilde{x}^*_{1nH}(z_0)} = \frac{(2n-1)^2}{n^2}.$$ ∎

Plugging Equation 8.41 into 8.38, we may obtain the payoff to an individual firm when every firm pitches in an x_0 learning effort and adopts the corresponding optimal stage 1 decision:

$$\tilde{g}_{1n}(x_0) = g_1(x_0, \tilde{x}^*_{1nL}(nx_0), \tilde{x}^*_{1nH}(nx_0)) = -x_0$$
$$+ [2\sqrt{HL} \cdot (1 + 2n\alpha x_0) + (L+H) \cdot (1 + 2n\alpha x_0 + 2n^2\alpha^2 x_0^2)]/(16n \cdot (1 + \alpha n x_0)^2).$$
(8.42)

Taking derivative, we find that

$$\frac{d\tilde{g}_{1n}(x_0)}{dx_0} = Q(A_1, n\alpha x_0),$$ (8.43)

where Q is defined in Equation 8.34, whereas A_1 is defined in Equation 8.35 but with $n = 1$. Suppose α, L, and H make $A_1 \geq 27/4$. Then, we may define $B_1 = w^*(A_1)$, as well as, C_1 so that

$$C_1 = n\alpha \cdot \left[\tilde{g}_{1n}\left(\frac{B_1}{n\alpha}\right) - \tilde{g}_{1n}(0) \right],$$ (8.44)

which by Equation 8.42, is the same as C_1 being defined through Equation 8.37 with $n = 1$. We can express the largest first-best individual-firm learning effort x^*_{1n0} in terms of these constants.

Proposition 8.8

It is true that

$$x^*_{1n0} = \begin{cases} 0, & \text{when } A_1 < 27/4, \text{ or } A_1 \geq 27/4 \text{ and } C_1 < 0, \\ B_1/(n\alpha), & \text{when } A_1 \geq 27/4 \text{ and } C_1 \geq 0. \end{cases}$$ ∎

8.4.4 Comparison between Sections 8.4.2 and 8.4.3

For $n = 2, 3,...$, let $z_{n0}^* = nx_{n0}^*$ be the total learning effort in equilibrium. Also, let $z_{10}^* = nx_{1n0}^*$ be the first-best total learning effort, which, according to Proposition 8.8, is independent of the number of firms n. Now we treat $n = 1, 2,...$ indiscriminately, with the understanding that $n = 1$ signifies the first-best case involving an arbitrary number of firms, while $n = 2, 3,...$ connotes the competitive case involving n firms.

From Propositions 8.7 and 8.8, it is clear that z_{n0}^* depends on L and H through the Ω-uncertainty indictor $\gamma = \sqrt{H} - \sqrt{L}$ only. We can further show that the total effort z_{n0}^* is decreasing in the number of decision makers n and will be encouraged by an increased level of return uncertainty γ; in addition, the total learning effect αz_{n0}^* increases with the effectiveness of learning α.

Theorem 8.3

For $n = 1,2,...$, A_n, B_n, and C_n are all decreasing in n, increasing in γ, and increasing in α. Consequently, z_{n0}^* is decreasing in n and increasing in γ; in addition, αz_{n0}^* is increasing in α. ∎

The aforementioned decrease of z_{n0}^* in n underscores the underlearning effect, that more intense competition leads to lower total investment in learning. With this, the individual learning effort x_{n0}^* will drop with n even faster. The above effect should be expected, as communal learning in some sense encourages "free riding." To help better understand the trends for A_n and C_n, we draw Figures 8.1 and 8.2. In Figure 8.1, we draw, in the (α, γ)-plane, iso-value curves $A_n = 27/4$ at different n values and gradients $(\partial A_n/\partial \alpha, \partial A_n/\partial \gamma)$ at different points of the curves. In Figure 8.2, we repeat the same for C_n.

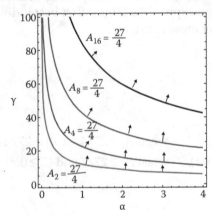

FIGURE 8.1
Iso-valued curves $A_n = 27/4$.

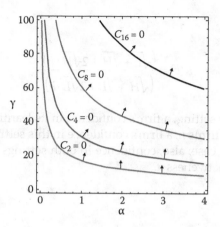

FIGURE 8.2
Iso-valued curves $C_n = 0$.

Let us consider the total stage 1 capacity investment levels. By Equations 8.31 and 8.41, we can combine the competitive and first-best cases to obtain the following: for $n = 1, 2, \ldots,$

$$\tilde{z}_{nL}^*(0) = \tilde{z}_{nH}^*(0) = \frac{(2n-1)^2 \cdot \left(\sqrt{L} + \sqrt{H}\right)^2}{16n^2}, \tag{8.45}$$

and

$$\begin{cases} \tilde{z}_{nL}^*(B_n/\alpha) = (2n-1)^2 \cdot \left(\sqrt{L} + \sqrt{H} + 2\sqrt{L} \cdot B_n\right)^2 / (16n^2 \cdot (1+B_n)^2), \\ \tilde{z}_{nH}^*(B_n/\alpha) = (2n-1)^2 \cdot \left(\sqrt{H} + \sqrt{L} + 2\sqrt{H} \cdot B_n\right)^2 / (16n^2 \cdot (1+B_n)^2). \end{cases} \tag{8.46}$$

We may define k_n^* for $n = 1,2,\ldots$ as a ratio:

$$k_n^* = \frac{\tilde{z}_{nH}^*(z_{n0}^*)}{\tilde{z}_{nL}^*(z_{n0}^*)}. \tag{8.47}$$

It measures the confidence a firm has in its acquired information about the market outlook.

By Propositions 8.7 and 8.8 and Equations 8.45 through 8.47, we have, for $n = 1, 2, \ldots,$

$$k_n^* = \begin{cases} 1, & \text{when } A_n < 27/4, \text{ or } A_n \geq 27/4 \text{ and } C_n < 0, \\ K(B_n), & \text{when } A_n \geq 27/4 \text{ and } C_n \geq 0, \end{cases} \tag{8.48}$$

where

$$K(b) = \frac{\left(\sqrt{L} + \sqrt{H} + 2\sqrt{H} \cdot b\right)^2}{\left(\sqrt{H} + \sqrt{L} + 2\sqrt{L} \cdot b\right)^2}. \tag{8.49}$$

In the competitive setting, a firm's confidence in its learning effect decreases with the number of firms n; a firm's confidence in this setting is always below that in the first-best case; also, confidence in both settings increases with the stage 0 learning effectiveness α.

Theorem 8.4

For $n = 1, 2,\ldots$, k_n^* is decreasing in n. Also, every k_n^* is increasing in α. ∎

8.5 A Potential Extension

In an extended learning model, we may still use the Ω defined through Equation 8.11 as firms' common belief of the size indicator before information acquisition. The relationship between each realized ω and the return function may still be described by Equation 8.15. However, this time we allow different firms to be able to acquire different information. On top of the random variable Θ, serving as firms' common observation through the relation in Equation 8.12, we introduce random variables Δ_1, Δ_2, ..., and Δ_n for the n firms. We let each Δ_i be a bivalued random variable ranging in $\{L, H\}$. When conditioned on $\Theta =$ some θ, the random variables $(\Delta_i|\theta)$ are independent of each other and $(\Omega|\theta)$. Also, there is a constant $b_i \in [0, 1]$ such that

$$\begin{cases} P[\Delta_i = L \,|\, \Theta = L] = (1 + b_i)/2 = P[\Delta_i = H \,|\, \Theta = H], \\ P[\Delta_i = L \,|\, \Theta = H] = (1 - b_i)/2 = P[\Delta_i = H \,|\, \Theta = L]. \end{cases} \tag{8.50}$$

Now, Θ serves as something intermediate between Ω and the Δ_i's. Firm i's final observation is the realization $\delta = L, H$ of the random variable Δ_i. The firm is to use this to make statistical inferences. By using Bayes' formula as well as other elementary probabilistic tools, it is easy to characterize the various random variables $(\Theta|\delta)$, $(\Delta_j|\delta)$ for $j \neq i$, and $(\Omega|\delta)$. We have

$$\begin{cases} P[\Theta = L \,|\, \Delta_i = L] = (1 + b_i)/2 = P[\Theta = H \,|\, \Delta_i = H], \\ P[\Theta = L \,|\, \Delta_i = H] = (1 - b_i)/2 = P[\Theta = H \,|\, \Delta_i = L]; \end{cases} \tag{8.51}$$

for $j \neq i$,

$$\begin{cases} P[\Delta_j = L \mid \Delta_i = L] = (1 + b_i b_j)/2 = P[\Delta_j = H \mid \Delta_i = H], \\ P[\Delta_j = L \mid \Delta_i = H] = (1 - b_i b_j)/2 = P[\Delta_j = H \mid \Delta_i = L]; \end{cases} \tag{8.52}$$

and,

$$\begin{cases} P[\Omega = L \mid \Delta_i = L] = (1 + a b_i)/2 = P[\Omega = H \mid \Delta_i = H], \\ P[\Omega = L \mid \Delta_i = H] = (1 - a b_i)/2 = P[\Omega = H \mid \Delta_i = L]. \end{cases} \tag{8.53}$$

Like in Section 8.3.1, we may suppose that there are stages 0 and 1. In stage 0, firms may invest to learn the market; in stage 1, they may participate in the investment game described in Section 8.2. Before stage 0, all firms believe that the return function follows $R_\Omega(x)$. Given stage 0 firm-effort vector $x_0 = (x_{0i} \mid i = 1,2,\ldots,n)$, we suppose that Θ is in the form of $\tilde{\Theta}(x_0)$ and each Δ_i is in the form of $\tilde{\Delta}(x_{0i}, x_{0,-i})$. Thus, $\tilde{\Theta}(x_0)$ reflects the communal learning effect and the $\tilde{\Delta}(x_{0i}, x_{0,-i})$'s reflect firms' individual take-aways. The true value of Ω will be revealed only after stage 1.

We suppose that the a used in Equation 8.12 is replaced by some function $\tilde{a}(x_0)$ and the b_i used in Equation 8.50 is replaced by some function $\tilde{b}(x_{0i}, x_{0,-i})$. With effort-dependent substitutions, we can achieve counterparts of the earlier Equations 8.51 through 8.53. For instance, for $\tilde{a}(\cdot)$, we may suppose that there is a positive constant α to satisfy Equation 8.14; for function $\tilde{b}(\cdot)$, we may suppose the existence of some positive constant to satisfy

$$\tilde{b}(x_{0i}, x_{0,-i}) = \frac{\beta x_{0i}}{1 + \beta x_{0i}}, \quad \forall i = 1,2,\ldots,n. \tag{8.54}$$

These function forms reflect that more can be learned through the exertion of greater efforts, and that the marginal return in learning decreases with effort levels. We have simplified the matter by letting $\tilde{b}(x_{0i}, x_{0,-i})$ depend on x_{0i} only. In the above, α again indicates the effectiveness of communal learning, while β indicates the strength of individual takeaways. At the extreme of $\beta = 0$, the final signal $\tilde{\Delta}(x_{0i}, x_{0,-i})$ will always be useless noise; at the other extreme of $\beta = +\infty$, $\tilde{\Delta}(x_{0i}, x_{0,-i})$ will be $\tilde{\Theta}(x_0)$ itself under the convention that $+\infty \cdot 0 = +\infty$. This last case is what we have just studied in Section 8.4.

We may use $x_i = (x_{0i}, x_{1i}(L), x_{1i}(H))$ to describe firm i's strategy. In it, x_{0i} is the firm's stage 0 learning effort, while $x_{1i}(\delta)$ is its stage 1 investment level when it has learned δ as the realization of $\tilde{\Delta}(x_{0i}, x_{0,-i})$. Let us use $f(x_i, x_{-i})$ to describe the average payoff to firm i, when it adopts policy $x_i = (x_{0i}, x_{1i}(L),$

$x_{1i}(H)$) while others have adopted policy profile $x_{-j} = ((x_{0j}, x_{1j}(L), x_{1j}(H))|j \neq i)$. We have

$$f(x_i, x_{-i}) = \sum_{\delta_i \in \{L,H\}} (1/2) \cdot \sum_{\delta_1 \in \{L,H\}} \cdots \sum_{\delta_{i-1} \in \{L,H\}} \cdot \sum_{\delta_{i+1} \in \{L,H\}} \cdots \sum_{\delta_n \in \{L,H\}}$$

$$\left\{ \Pi_{j \neq i} p(\delta_j | \delta_i; x_{0i}, x_{0j}) \times \sum_{\omega \in \{L,H\}} q(\omega | \delta_i; x_{0i}, x_{0,-i}) \times [(x_{1i}(\delta_i)/(x_{1i}(\delta_i)) \right.$$

$$\left. + \sum_{j \neq i} x_{1j}(\delta_j))) \cdot \omega \cdot r((x_{1i}(\delta_i) + \sum_{j \neq i} x_{1j}(\delta_j))/\omega) - x_{0i} - x_{1i}(\delta_i)] \right\}, \qquad (8.55)$$

where, according to effort-dependent versions of Equations 8.52 and 8.54,

$$p(\delta_j | \delta_i; x_{0i}, x_{0j}) = \begin{cases} (1 + \beta x_{0i} + \beta x_{0j} + 2\beta^2 x_{0i} x_{0j})/(2 \cdot (1 + \beta x_{0i}) \cdot (1 + \beta x_{0j})), \\ \qquad \text{when } \delta_i = \delta_j, \\ (1 + \beta x_{0i} + \beta x_{0j})/(2 \cdot (1 + \beta x_{0i}) \cdot (1 + \beta x_{0j})), \\ \qquad \text{when } \delta_i \neq \delta_j, \end{cases}$$

$$(8.56)$$

and according to effort-dependent versions of Equations 8.14, 8.53, and 8.54,

$$q(\omega | \delta_i; x_{0i}, x_{0,-i}) = \begin{cases} \left(1 + (\alpha + \beta)x_{0i} + \alpha \sum_{j \neq i} x_{0j} + 2\alpha\beta x_{0i} \cdot \left(x_{0i} + \sum_{j \neq i} x_{0j} \right) \right) / \\ \left(2 \cdot \left(1 + \alpha x_{0i} + \alpha \sum_{j \neq i} x_{0j} \right) \cdot (1 + \beta x_{0i}) \right), \quad \text{when } \delta_i = \omega, \\ \left(1 + (\alpha + \beta)x_{0i} + \alpha \sum_{j \neq i} x_{0j} \right) / \\ \left(2 \cdot \left(1 + \alpha x_{0i} + \alpha \sum_{j \neq i} x_{0j} \right) \cdot (1 + \beta x_{0i}) \right), \quad \text{when } \delta_i \neq \omega. \end{cases}$$

$$(8.57)$$

Inside Equation 8.55, the 1/2 is the chance for firm i's observation Δ_i to be either L or H, $p(\delta_j|\delta_i; x_{0i}, x_{0j})$ is the conditional probability $P[\Delta_j = \delta_j|\Delta_i = \delta_i]$ when firms' learning-effort vector is $x_0 = (x_{0i}|i = 1,2,\dots,n)$, and $q(\omega|\delta_i; x_{0i}, x_{0,-i})$ is the conditional probability $P[\Omega = \omega|\Delta_i = \delta_i]$ under the same learning-effort vector. Note that firm i's decision is only dependent on δ_i, whereas its payoff is dependent on the actual Ω-realization ω.

When $\beta < +\infty$, different firms receive different signals, and Equation 8.55 is much more difficult to analyze than its counterpart (Equation 8.17) at $\beta = +\infty$. Outcomes will diverge depending on whether or not firms share these signals. It turns out that the concerned problem can become entangled even when we assume that there are $n = 2$ firms and that the return function $r(x) = \sqrt{x}$. Our preliminary analysis is numerical in nature, from which we may draw some conclusion on conditions that favor information sharing among firms. But more extensive analysis is still needed.

8.6 Concluding Remarks

We formulated a capacity investment game in which identical firms contest for market shares and watch out for their investment returns at the same time. Overcapacity appears as a natural outcome of this game. By introducing uncertainty to the market size and adding a stage 0 of competitive information gathering to the game, we enabled the investigation of information and learning in a competitive setting. Our theoretical analysis confirmed the severe underlearning effect when incentives exist for "free riding."

Extending on the learning framework in Section 8.3, we may further model, as we did in Section 8.5, the case where different firms receive different signals. Outcomes will diverge depending on whether or not firms share these signals. We may then study whether or not information sharing brings in additional benefits. So far we have found that analysis would become numerically entangled even for two firms. But we hope that, in future research, some minor adjustments to the current framework could pave the way for more major insights.

Appendix

A. Proof of Proposition 8.1

It all hinges on h_n. Note that

$$h'_n(x) = r''(nx) + \frac{n-1}{nx} \cdot \left[r'(nx) - \frac{r(nx)}{nx} \right] - c''(x), \qquad (8.58)$$

which, by (r2), (r02), and (c2), is strictly negative. By l'Hôpital's rule,

$$h_n(0^+) = r'(0^+) - c'(0^+), \qquad (8.59)$$

which is strictly positive by (cr). Also, by (r3), (c1), and (c2), we know that $h_n(x) < 0$ will occur when x is large enough. Therefore, x_n^* is in existence and unique.

B. Proof of Proposition 8.2

By (r02), (c2), and (8), we know that $i_n(z)$ is increasing in n at any fixed z. Thus, z_n^* is increasing in n, as it is the unique root of the decreasing function $i_n(z)$.

C. Proof of Proposition 8.3

By (c2) and (10), we know that $h_{1n}(z)$ is increasing in n at any fixed z. Thus, z_{1n}^* is increasing in n, as it the unique root of the decreasing function $h_{1n}(z)$. On the other hand, $x_{1n}^* = z_{1n}^*/n$ is the unique root for the decreasing function $r'(nx) - c'(x)$. By (r2), we know that the function is decreasing in n. Thus, we know that x_{1n}^* is decreasing in n.

D. Proof of Theorem 8.1

We clearly have $z_1^* = z_{11}^*$. Let us focus on the case where $n \geq 2$. Comparing Equation 8.8 with Equation 8.10, we have

$$i_n(z) - i_{1n}(z) = \frac{n-1}{n} \cdot \left[\frac{r(z)}{z} - r'(z) \right], \tag{8.60}$$

which is strictly positive by (r02). With z_n^* and z_{1n}^* being roots, respectively, of the functions i_n and i_{1n}, we therefore have $z_n^* > z_{1n}^*$.

E. Proof of Lemma 8.1

By Equation 8.21, we have

$$
\begin{aligned}
\partial g_L'(z_0, x_L, y_L)/\partial x_L = &[(1+2\alpha z_0)/(4+4\alpha z_0)] \times [(x_L/(L \cdot (x_L+y_L))) \cdot r''((x_L+y_L)/L) \\
&+ (2y_L/(x_L+y_L)^2) \cdot r'((x_L+y_L)/L) \\
&- (2y_L/(x_L+y_L)^3) \cdot L \cdot r((x_L+y_L)/L)] \\
&+ [1/(4+4\alpha z_0)] \times [(x_L/(H \cdot (x_L+y_L))) \cdot r''((x_L+y_L)/H) \\
&+ (2y_L/(x_L+y_L)^2) \cdot r'((x_L+y_L)/H) \\
&- (2y_L/(x_L+y_L)^3) \cdot H \cdot r((x_L+y_L)/H)].
\end{aligned}
\tag{8.61}
$$

That is, we have

$$\frac{\partial g_L'(z_0, x_L, y_L)}{\partial x_L} = S_1 \cdot (Q_1 + S_2 T_1) + S_3 \cdot (Q_2 + S_2 T_2), \qquad (8.62)$$

where

$$\begin{cases} Q_1 = (x_L/(L \cdot (x_L + y_L))) \times r''((x_L + y_L)/L), \\ Q_2 = (x_L/(H \cdot (x_L + y_L))) \times r''((x_L + y_L)/H), \end{cases} \qquad (8.63)$$

$$\begin{cases} S_1 = (1 + 2\alpha z_0)/(4 + 4\alpha z_0), \quad S_2 = 2y_L/(x_L + y_L)^2, \\ S_3 = 1/(4 + 4\alpha z_0), \end{cases} \qquad (8.64)$$

and

$$\begin{cases} T_1 = r'((x_L + y_L)/L) - r((x_L + y_L)/L)/((x_L + y_L)/L), \\ T_2 = r'((x_L + y_L)/H) - r((x_L + y_L)/H)/((x_L + y_L)/H). \end{cases} \qquad (8.65)$$

By (r2), we have $Q_1 < 0$ and $Q_2 < 0$; S_1, S_2, and S_3 are all apparently strictly positive; also, by (r02), we have $T_1 < 0$ and $T_2 < 0$. Therefore, we have $\partial g_L'(z_0, x_L, y_L)/\partial x_L < 0$. Symmetrically, we can show that $\partial g_H'(z_0, x_H, y_H)/\partial x_H < 0$.

F. Proof of Proposition 8.4

Because $j_n(x)$ is strictly decreasing in x, we know from Equation 8.23 that $h_{nL}(z_0, x_L)$ is strictly decreasing in x_L. By l'Hôpital's rule, we know that $j_n(0^+) = r'(0^+)$, and hence

$$h_{nL}(z_0, 0^+) = \frac{r'(0^+) - 1}{2}, \qquad (8.66)$$

which is strictly positive by (cr). Also, by (r3), we know that $h_{nL}(z_0, x_L) < 0$ will occur when x_L is large enough. Therefore, the root $\tilde{x}_{nL}^*(z_0)$ is in existence and unique. The result on $\tilde{x}_{nH}^*(z_0)$ can be achieved symmetrically.

G. Proof of Proposition 8.5

From Equations 8.23 and 8.25, we see that

$$h_{nL}(0, x) = h_{nH}(0, x) = -\frac{1}{2} + \frac{1}{4} \cdot \left[j_n\left(\frac{nx}{L}\right) + j_n\left(\frac{nx}{H}\right) \right]. \qquad (8.67)$$

Therefore, the unique roots $x_{nL}^*(0)$ and $x_{nH}^*(0)$ of the preceding two identical functions should be equal to each other. Rewriting Equations 8.23 and 8.25, we note that

$$h_{nL}(z_0, x_L) = -\frac{1}{2} + \frac{1}{2} \cdot j_n\left(\frac{nx_L}{L}\right) + \frac{1}{4 + 4\alpha z_0} \cdot \left[j_n\left(\frac{nx_L}{H}\right) - j_n\left(\frac{nx_L}{L}\right) \right], \quad (8.68)$$

and

$$h_{nH}(z_0, x_H) = -\frac{1}{2} + \frac{1}{2} \cdot j_n\left(\frac{nx_H}{H}\right) + \frac{1}{4 + 4\alpha z_0} \cdot \left[j_n\left(\frac{nx_H}{L}\right) - j_n\left(\frac{nx_H}{H}\right) \right]. \quad (8.69)$$

Because $j_n(x)$ is strictly decreasing in x and $H > L > 0$, we know that $j_n(nx_L/H) > j_n(nx_L/L)$ and $j_n(nx_H/L) < j_n(nx_H/H)$. Hence, $h_{nL}(z_0, x_L)$ is decreasing in z_0 and $h_{nH}(z_0, x_H)$ is increasing in z_0. Being roots of decreasing functions $h_{nL}(z_0, \cdot)$ and $h_{nH}(z_0, \cdot)$, we therefore know that $\tilde{x}_{nL}^*(z_0)$ is decreasing in z_0 and that $\tilde{x}_{nH}^*(z_0)$ is increasing in z_0.

H. Proof of Lemma 8.2

First, we always have $Q(a, 0) = -1$. Now we analyze $Q(a, \cdot)$ for $a \geq 0$. Note that

$$\frac{\partial Q(a, w)}{\partial w} = \frac{a \cdot (1 - 2w)}{(1 + w)^4}, \quad (8.70)$$

and

$$\frac{\partial^2 Q(a, w)}{\partial w^2} = \frac{6a \cdot (w - 1)}{(1 + w)^5}. \quad (8.71)$$

It is easy to see that $Q(a, \cdot)$ has a local maximum at $1/2$. We have

$$Q\left(a, \frac{1}{2}\right) = \frac{4a}{27} - 1. \quad (8.72)$$

Therefore, when $a \in [0, 27/4)$, the function $Q(a, \cdot)$ is never above 0 for $w \in R^+$; when $a \in [27/4, +\infty)$, however, $Q(a, \cdot)$ is above 0 in an interval containing $1/2$, whose left- and right-end points are the two desired roots $w^0(a)$ and $w^*(a)$.

I. Proof of Proposition 8.7

We know the following from Lemma 8.2: When $A_n \in [0, 27/4)$, the function $Q(A_n, \cdot)$ is never above 0 for $w \in R^+$, and hence $\tilde{g}_n(x_0, y_0)$ is never increasing in x_0; when $A_n \in [27/4, +\infty)$, however, $Q(A_n, \cdot)$ is above 0 in the interval $[w^0(A_n), w^*(A_n)]$ containing $1/2$.

Now, we know that $\tilde{g}_n(x_0, y_0)$ is decreasing in x_0 when $\alpha(x_0 + y_0) \in [0, w^0(A_n))$, increasing in x_0 when $\alpha(x_0 + y_0) \in [w^0(A_n), w^*(A_n))$, and decreasing in x_0 again when $\alpha(x_0 + y_0) \in [w^*(A_n), +\infty)$. When $A_n \geq 27/4$, we may define D_n so that

$$D_n = \alpha \cdot \left[\tilde{g}_n\left(\frac{B_n}{\alpha}, 0 \right) - \tilde{g}_n(0,0) \right]. \tag{8.73}$$

We can show that $(D_n \geq 0) \Rightarrow (C_n \geq 0)$, and hence $(C_n < 0) \Rightarrow (D_n < 0)$. By Equations 8.32 and 8.73, we have

$$D_n = \frac{A_n B_n^2}{2 \cdot (1 + B_n)^2} - B_n, \tag{8.74}$$

which, because $Q(A_n, B_n) = Q(A_n, w^*(A_n)) = 0$, leads to

$$D_n = \frac{B_n \cdot (B_n - 1)}{2}. \tag{8.75}$$

By Equation 8.75 and the fact that $B_n \geq 1/2$, we have $(D_n \geq 0) \Rightarrow (B_n \geq 1)$. But by Equation 8.37, $B_n \geq 1$ leads to $C_n \geq 0$.

When $A_n < 27/4$, as $\tilde{g}_n(\cdot, 0)$ is decreasing, $\tilde{g}_n(0,0)$ is greater than $\tilde{g}_n(x_0, 0)$ for any $x_0 \in R^+$. Also, $\tilde{g}_n(x_0, (n-1)x_0)$ is smaller than $\tilde{g}_n(0, (n-1)x_0)$ for any $x_0 \in (0, +\infty)$, because $\tilde{g}_n(\cdot, (n-1)x_0)$ is decreasing too. Hence, we have $x_{n0}^* = 0$.

When $A_n \geq 27/4$ and C_n as defined in Equation 8.36 is positive, it follows that $\tilde{g}_n(B_n/(n\alpha), (n-1)B_n/(n\alpha))$ is greater than $\tilde{g}_n(0, (n-1)B_n/(n\alpha))$, the only other local maximum of $\tilde{g}_n(\cdot, (n-1)B_n/(n\alpha))$; thus, $\tilde{g}_n(B_n/(n\alpha), (n-1)B_n/(n\alpha))$ is greater than $\tilde{g}_n(x_0, (n-1)B_n/(n\alpha))$ for any $x_0 \in R^+$. Also, $\tilde{g}_n(x_0, (n-1)x_0)$ is smaller than $\tilde{g}_n(B_n/(n\alpha), (n-1)x_0)$ for any $x_0 \in (B_n/(n\alpha), +\infty)$, since $\tilde{g}_n(\cdot, (n-1)x_0)$ is decreasing when the argument is above $B_n/(n\alpha)$. Hence, we have $x_{n0}^* = B_n/(n\alpha)$.

When $A_n \geq 27/4$ but $C_n < 0$, we have $D_n < 0$ according to the above. Hence, it follows that $\tilde{g}_n(0,0)$ is greater than $\tilde{g}_n(B_n/\alpha, 0)$, the only other local maximum of $\tilde{g}_n(\cdot, 0)$; thus, $\tilde{g}_n(0,0)$ is greater than $\tilde{g}_n(x_0, 0)$ for any $x_0 \in R^+$. Also, $\tilde{g}_n(x_0, (n-1)x_0)$ is smaller than $\tilde{g}_n((B_n/\alpha - (n-1)x_0) \vee 0, (n-1)x_0)$ for any $x_0 \in (0, +\infty)$. Hence, we have $x_{n0}^* = 0$.

J. Proof of Proposition 8.8

We know the following from Lemma 8.2: When $A_1 \in [0, 27/4)$, the function $Q(A_1, \cdot)$ is never above 0 for $w \in R^+$, and hence $\tilde{g}_{1n}(x_0)$ is never increasing in x_0; when $A_1 \in [27/4, +\infty)$, however, $Q(A_1, \cdot)$ is above 0 in the interval $[w^0(A_1), w^*(A_1)]$ containing 1/2. In the latter case, $\tilde{g}_{1n}(\cdot)$ has two local maximums, 0 and $B_1/(n\alpha)$. Whether $x_{1n0}^* = 0$ or $x_{1n0}^* = B_1/(n\alpha)$ depends solely on whether or not $\tilde{g}_{1n}(0) - \tilde{g}_{1n}(B_1/(n\alpha)) > 0$, which, according to Equation 8.44, is the same as $C_1 < 0$.

K. Proof of Theorem 8.3

From Equation 8.35, it is clear that A_n is decreasing in n, and increasing in γ and α. For $a \in [27/4, +\infty)$, we may take derivative of a on the equation $Q(a, w^*(a)) = 0$ while in consultation with Equation 8.34, to obtain

$$\frac{dw^*(a)}{da} = \frac{w^*(a)}{3(1+w^*(a))^2 - a} = \frac{(w^*(a))^2}{2(w^*(a))^3 + 3(w^*(a))^2 - 1}, \tag{8.76}$$

which is positive since $w^*(a) \geq 1/2$. Hence, just because A_n is so, $B_n = w^*(A_n)$ is decreasing in n, and increasing in γ and α.

By Equation 8.37, we have

$$C_n = G(n, B_n), \tag{8.77}$$

where

$$G(k,b) = \frac{(2k-2) \cdot b^3 + kb \cdot (b-1)}{2 \cdot (kb - b + k)^2}. \tag{8.78}$$

Taking derivatives, we have

$$\frac{\partial G(k,b)}{\partial k} = \frac{b^2 \cdot [1 - (k-1) \cdot (2b^2 + 3b)] + kb}{D(k,b)}, \tag{8.79}$$

and

$$\frac{\partial G(k,b)}{\partial b} = \frac{N(k,b)}{D(k,b)}, \tag{8.80}$$

where

$$N(k, b) = 2(k - 1)^2 \cdot b^3 + 6k(k - 1) \cdot b^2 + k(3k - 1) \cdot b - k^2, \quad (8.81)$$

and

$$D(k, b) = 2 \cdot (kb - b + k)^3. \quad (8.82)$$

From Equation 8.35, we have, when treating A_n as a function of n, γ, and α,

$$\frac{\partial A_n}{\partial n} = -\frac{(4n - 3)\alpha\gamma^2}{8n^4} = -\frac{(4n - 3) \cdot A_n}{2n^2 - n}. \quad (8.83)$$

In view of $Q(A_n, B_n) = 0$, this leads to

$$\frac{\partial A_n}{\partial n} = -\frac{(4n - 3) \cdot (1 + B_n)^3}{(2n^2 - n) \cdot B_n}. \quad (8.84)$$

Combining the preceding while treating C_n as a function of n, γ, and α, we obtain

$$\partial C_n / \partial n = \partial G(k, b) / \partial k \big|_{k=n, b=B_n} + \partial G(k, b) / \partial b \big|_{k=n, b=B_n}$$
$$\times \, dw^*(a) / da \big|_{w^*(a) = B_n} \times \partial A_n / \partial n = -(n - 1) \cdot B_n \cdot (1 + B_n)$$
$$\cdot J(n, B_n) / [(2n^2 - n) \cdot (nB_n - B_n + n)^3 \cdot (2B_n - 1)], \quad (8.85)$$

where

$$J(k, b) = 3b^3 - b(1 + 9b(1 + b)) \cdot k + (1 + b)(8b^2 + 4b - 1) \cdot k^2. \quad (8.86)$$

Note that $J(\cdot, b)$ is a quadratic function with the minimum achieved at $k_0(b) = 1 - (7b^3 + 15b^2 + 5b - 2)/(2 \cdot (8b^3 + 12b^2 + 3b - 1))$, which is below 1 for $b \geq 1/2$. Thus, for $k \geq 1$ and $b \geq 1/2$, we have $J(k, b) \geq J(1, b) = 2b^3 + 3b^2 + 2b - 1 \geq 0$. Thus, we know that $\partial C_n / \partial n \leq 0$, and hence C_n is decreasing in n.

For $k \geq 1$ and $b \geq 1/2$, $D(k, b)$ as defined in Equation 8.82 is clearly positive. Also, from Equation 8.81, we have

$$N(k, b) \geq \frac{1}{4} \cdot (k - 1)^2 + \frac{3}{2} \cdot k(k - 1) + \frac{1}{2} \cdot k(3k - 1) - k^2 = \frac{(9k - 1)(k - 1)}{4} \geq 0. \quad (8.87)$$

By (80), we may achieve from this the positivity of $\partial G(k, b)/\partial b$. Hence, just because B_n is so, C_n is increasing in γ and α.

The remaining results are simple consequences of the above trends of A_n, B_n, and C_n, as well as Propositions 8.7 and 8.8.

We now know that the strictly positive z_{n0}^*-value of B_n/α is decreasing in n. Suppose n has been increased. Then, it will be less likely for $A_n \geq 27/4$ and $C_n \geq 0$ to occur, and hence less likely for z_{n0}^* to assume the strictly positive value. Therefore, z_{n0}^* is decreasing in n.

We also know that the strictly positive z_{n0}^*-value of B_n/α is increasing in γ. Suppose γ has been increased. Then, it will be more likely for $A_n \geq 27/4$ and $C_n \geq 0$ to occur, and hence more likely for z_{n0}^* to assume the strictly positive value. Therefore, z_{n0}^* is increasing in γ.

Moreover, we know that the strictly positive αz_{n0}^*-value of B_n is increasing in α. Suppose α has been increased. Then, it will be more likely for $A_n \geq 27/4$ and $C_n \geq 0$ to occur, and hence more likely for αz_{n0}^* to assume the strictly positive value. Therefore, αz_{n0}^* is increasing in α.

L. Proof of Theorem 8.4

From Theorem 8.3, we know that A_n and C_n are all decreasing in n. Hence, from Equation 8.48, we have

$$k_n^*/k_{n+1}^* = \begin{cases} 1, & \text{when } A_n < 27/4, \text{ or } A_n \geq 27/4 \text{ and } C_n < 0, \\ K(B_n)/K(B_{n+1}), & \text{when } A_{n+1} \geq 27/4 \text{ and } C_{n+1} \geq 0, \\ K(B_n), & \text{in all other cases.} \end{cases}$$

(8.88)

From Equation 8.49, it is easy to check that, as b increases from $1/2$ to $+\infty$, $K(b)$ increases from $\left(2\sqrt{H}+\sqrt{L}\right)^2/\left(\sqrt{H}+2\sqrt{L}\right)^2$ to H/L. By Theorem 8.3 again, we know that B_n is decreasing in n. Combining these together, we will have $k_n^*/k_{n+1}^* \geq 1$ in all situations.

The increase of k_n^* in α is a simple consequence of Equation 8.48, the above fact about the $K(b)$ function, and the increase of A_n, B_n, and C_n in α as stipulated in Theorem 8.3.

References

Acemoglu, D., K. Bimpikis, and A. Ozdaglar. Price and capacity competition. *Games and Economic Behavior*, 66(1): 1–26, 2009.

Aghion, P., M. P. Espinosa, and B. Jullien. Dynamic duopoly with learning through market experimentation. *Economic Theory*, 3(3): 517–539, 1993.

Anupindi, R. and L. Jiang. Capacity investment under postponement strategies, market competition, and demand uncertainty. *Management Science*, 54(11): 1876–1890, 2008.

Bensoussan, A., M. Cakanyildirim, and S. P. Sethi. A multi-period news vendor problem with partially observed demands. *Mathematics of Operations Research*, 32(2): 322–344, 2007.

Blackwell, D. Comparison of experiments. *Proceedings of the Second Berkeley Symposium on Mathematical Statistics and Probability* (pp. 93–102). Berkeley: University of California Press, 1951.

Bulow, J., J. Geanakoplos, and P. Klemperer. Holding idle capacity to deter entry. *Economic Journal*, 95(377): 178–182, 1985.

Burnetas, A. and S. M. Gilbert. Future capacity procurements under unknown demand and increasing costs. *Management Science*, 47(7): 979–992, 2001.

Burns, S. LG Philips LCD announces $335m loss. *Australian PC Authority*, October 13, 2006. Avalable at http://www.pcauthority.com.au.

Cachon, G. P. Supply chain coordination with contracts. In S. Graves and T. de Kok (eds.), *Handbooks in Operations Research and Management Science: Supply Chain Management*. Amsterdam: North Holland, 2003.

Chen, L. and E. Plambeck. Dynamic inventory management with learning about the demand distribution and substitution probability. *Manufacturing and Service Operations Management*, 10(2): 236–256, 2008.

Davidson, C. and R. Deneckere. Long-run competition in capacity, short-run competition in price, and the Cournot model. *Rand Journal of Economics*, 17(3): 404–415, 1986.

Deneckere, R., H. Marvel, and J. Peck. Demand uncertainty and price maintenance: Markdowns as destructive competition. *American Economic Review*, 87(4): 619–641, 1997.

Dixit, H. The role of investment in entry deterrence. *Economic Journal*, 90(1): 95–106, 1980.

Dressler, A. Autos: Overinvestment and overcapacity are causing concern. *fdimagazine.com*, October 20, 2004. Avalable at http://www.fdimagazine.com/news.

Ihlwan, M. Samsung's chip business: What happened? *Business Week*, June 15, 2007.

Kardos, D., J. Flowers, and I. S. Nam. Business technology: Prices tumble for memory products. *The Wall Street Journal*, April 1, B7, 2008.

Kreps, D. M. and J. A. Scheinkman. Quantity precommitment and Bertrand competition yield Cournot outcomes. *The Bell Journal of Economics*, 14(2): 326–337, 1983.

Lariviere, M. and E. Porteus. Stalking information: Bayesian inventory management with unobserved lost sales. *Management Science*, 45(3): 346–363, 1999.

Lehmann, E. L. Comparing location experiments. *Annals of Statistics*, 16(2): 521–533, 1988.

Lippman, S. A. and K. F. McCardle. The competitive newsboy. *Operations Research*, 45(1): 54–65, 1997.

Mahajan, S. and G. van Ryzin. Inventory competition under dynamic consumer choice. *Operations Research*, 49(5): 646–657, 2001.

Rob, R. Learning and capacity expansion under demand uncertainty. *Review of Economic Studies*, 58(4): 655–675, 1991.

Shin, H. D. and T. Tunca. The effect of competition on demand forecast investments and supply chain coordination. Working Paper, Northwestern University, 2009.

9

A Multiperiod Multiclass High-Speed Rail Passenger Revenue Management Problem

Ying Qin, Zhe Liang, Wanpracha Art Chaovalitwongse, and Shaozhong Xi

CONTENTS

ABSTRACT   In this chapter, we study a multiperiod multiclass rail passenger revenue management (MPMC-RPRM) problem. In MPMC-RPRM, we assume that the unsatisfied demand from a previous period can be recaptured by the later period, and the unsatisfied demand from a class can be recaptured by other classes. To formulate the MPMC-RPRM, we first propose a basic model (BM). Because there are a large number of binary variables and big-M constraints in the basic model, it is very time consuming to obtain the optimal solutions for the real-life large-scale problems. Therefore, we propose a two-stage heuristic to solve the problem. In the first stage, we decompose each trip into a number of consecutive legs. Instead of maximizing the total revenue from all origin–destination pairs, we maximize the total revenue from all legs. The result of the first-stage heuristic provides a set of trains needed. In the second stage of the heuristic, we maximize the total revenue from all the origin–destination pairs on a restricted BM model, in which only the set of trains that have been fixed in the first stage can be used. The computational results show that the two-stage heuristic provides a better solution than the basic model in a much shorter time. We also extend our model and solution approach with some real-life considerations.

KEY WORDS: *decomposition heuristic, integer programming, passenger revenue management, rail transportation, scheduling.*

9.1 Introduction

Half a century ago, on October 1, 1964, the first modern high-speed rail, Shinkansen, began to service passengers between Tokyo and Osaka with a top speed of 210 km/h (130 mph). Since then, the development of high-speed rail has brought enormous economic growth and spread prosperity all over the world. Today, high-speed rails (HSRs) are being operated in many countries and areas, including the Shinkansen in Japan, Eurostar, Acela Express in the Northeast Corridor of the United States, and China Railways Highspeed (CRH), to name but a few. In recent years, the demand for high-speed rail has been growing, partially because other modes of major transport have faced increasing challenges such as highway congestion, cancelations and delays of flights, increasing cost of fuel, and so forth. Also, rail transportation is more energy and cost efficient than cars and airplanes. Bureau of Transportation Statistics data shows that the passenger rail is 30%–50% more energy efficient than cars and airplane. Despite the rapid growth of the HSRs, financial profitability remains a serious challenge for almost all HSRs. A recent report by the Reason Foundation showed that only two HSRs in the world are profitable: Paris–Lyon in France and Shinkansen between Tokyo and Osaka in Japan (Feigenbaum, 2013). Therefore, it is critical to optimize the revenue of the day-to-day operations to ensure the sustainability of HSR. However, unlike airline revenue management, for which one can easily find copious literature, there is very little research that focuses on rail passenger revenue management (RPRM).

In this chapter, we focus on a multiperiod multiclass RPRM problem (MPMC-RPRM). The main difference between our problem and the others in the

RPRM literature is that we assume a portion of the unsatisfied demand from the previous period contributes to the demand of the next period. Similarly, we assume a portion of the unsatisfied demand from one seat class can be captured by the adjacent seat classes. By considering the spilled demands between periods and seat classes, we provide a more realistic description of passenger booking behavior. We then present a complete mathematical programming formulation to solve the problem. Because there are a large number of binary variables and big-M constraints in the proposed model, it is impossible to solve real-life large-scale problems in a reasonable time. Therefore, we propose a decomposition-based two-stage heuristic to solve the problem efficiently.

The remainder of the chapter is organized as follows. In Section 9.2, we give a brief review of the RPRM problems. In Section 9.3, we define the MPMC-RPRM formally and propose a basic mathematical model for the problem. In Section 9.4, we present a two-stage heuristic to solve the proposed models efficiently. In Section 9.5, we discuss some model extensions for operational considerations. In Section 9.6, we provide the computational experience for the set of the test cases. Section 9.10 concludes the chapter.

9.2 Background

In this section, we provide a detailed review of the RPRM problem. We also provide background information on a particular HSR, Chinese CRH between Beijing and Shanghai, which motivates our research in this chapter.

9.2.1 Rail Passenger Revenue Management

The RPRM is closely related with the airline revenue management (ARM), where the airline companies try to maximize the profits from a fixed perishable resource, that is, airline seats. Extensive literature is available on ARM, and one can refer to McGill and Van Ryzin (1999), Talluri and Van Ryzin (2004), Bertsimas and de Boer (2005), Cooper and Gupta (2006), Cooper and de Mello (2007), Barnhart et al. (2009), Zhang et al. (2010), Lan et al. (2011), and Aydin et al. (2013) for a comprehensive review. On the other hand, only a very limited number of papers available for RPRM. Ciancimino et al. (1999) presented a deterministic model and a probabilistic model for a multileg single-fare railway seat allocation problem. Their computational study showed increasing revenue when applying both models. Bharill and Rangaraj (2008) developed a model to estimate the cross-price demand elasticity when the ticket fare and cancelation cost change for a premium segment of Indian Railways. The authors then used the demand elasticity to estimate the demand and to analyze the existing pricing strategies. You (2008) proposed a constrained nonlinear integer programming model for a two-fare multileg seat allocation problem. The author also developed a heuristic that hybridizes the mathematical approach and particle swarm optimization framework to find the booking limits for the problem. Armstrong and Meissner (2010) provided a comprehensive review on both passenger and freighter rail revenue management. The authors categorized the RPRM literature by seven problem properties such as class fare, number of legs, number of services, pricing strategies, and so on. The authors also pointed out two major differences between the RPRM problem and the traditional ARM problem. The first is that no overbooking is considered in the RPRM because normally the load of a train is less than 100%, and a small number of customers are still allowed to be on board even after all the seats have been occupied. The second is that the legs in RPRM are highly correlated because most rail trips contain multiple adjacent legs, whereas in ARM, the majority of trips contain only a single leg or two connecting legs. Dutta and Ghosh (2012) presented a linear mathematical model for a multiperiod multiclass multileg seat allocation problem. However, their multiperiod multiclass model can be decomposed directly into multiple single-period single-class models because there is no interrelationship between any two planning periods or any two classes in terms of demands and ticketing decisions. Cirillo and Hetrakul (2011) presented an optimization model for a multiperiod single-leg dynamic pricing problem. In their model, the authors used a multinomial logit model to estimate the passenger choice of booking day, and a least squares regression model for the demand of each market. Hetrakul and Cirillo (2013) used several statistic models, including multinomial logit, mixed logit, and latent class models to estimate the demand of online passenger advanced booking. Their results shown that fare price, advance booking (number of days before departure), and departure day of week can be used to determine the demand of advance online booking. They suggested that the proposed demand estimation models can be used to support the railway revenue management policies such as pricing and seat allocation. Recently, Crevier et al. (2012) proposed solving the operation planning and revenue management simultaneously for rail freight transportation using a bilevel mathematical formulation. Two pricing policies are proposed and their impacts on the model are analyzed.

9.2.2 High-Speed Rail in China

In the last seven years, China has built more than 13,000 kilometers of high-speed rail and plans to extend it to 20,000 kilometers by the end of 2015. In 2013, the daily ridership of China CRH was more than 1.45 million (China Railway Corporation, 2014), and in a recent report from the *New York Times*, it was predicted that "China's high-speed rail network will handle more passengers by early 2014 than the 54 million people a month who board domestic flights in the United States" (Bradsher, 2013). One of the busiest CRHs is Beijing–Shanghai high-speed railway (BSHR), which is 1318 kilometers

(819 miles) long and passes 23 cities along the line (Wikipedia, 2014), as shown in Figure 9.1.

The travel time from Beijing to Shanghai is 5.5 hours on average, ranging from 4 hours 48 minutes to 6 hours depending on the number of stops along the line. In year 2013, the average daily ridership of BSHR is 230,000, and the average occupancy rate of the train is 76.9% (China.com, 2014).

Currently, fixed fare policy is operated for all the passenger services in China. Passengers can buy tickets online or directly from the railway stations

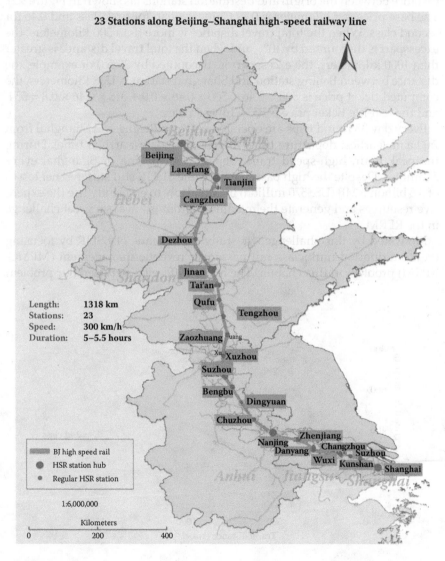

FIGURE 9.1
Beijing–Shanghai high-speed railway.

up to 2 weeks before the departure. Similarly to air tickets, each train ticket is for a particular seat in the train. For the BSHR, three classes are available: business class, first class, and second class. Two fleets are operated for BSHR, mainly differing in the number of seats available. One of the fleets contains 13 business class seats, 85 first class seats, and 854 second class seats, and the other fleet contains 26 business class seats, 123 first class seats, and 776 second class seats.

The ticket price of all CRH is determined mainly on the basis of the travel distance between the origin and destination stations (as shown in Figure 9.2). The base price per kilometer is 0.79 Chinese yuan for first class and 0.46 for second class. When the total travel distance is more than 500 kilometers, the excesspart is discounted by 10%, and when the total travel distance is greater than 1000 kilometers, the excesspart is discounted by 20%. For example, the distance between Beijing station and Shanghai station is 1318 kilometers, the computed ticket price is $500 \times 0.46 + 500 \times 0.46 \times 0.9 + 318 \times 0.46 \times 0.8 = 554$, and the real-life ticket price is 553 Chinese yuan.

Every day 33 round trips are operated between Beijing and Shanghai from 7:00 a.m. (earliest departure time) to 11:30 p.m. (latest arrival time). During the peak hour, high-speed trains depart from Beijing or Shanghai every 5 minutes. Despite the high passenger volume, BSHR still suffered net losses of 3.5 billion RMB (US$570 million) in 2012. How to efficiently use the expensive resources and generate the maximum profit is always a great challenge in the RPRM.

Motivated by this challenge, we study the revenue of BSHR by focusing on a multiperiod multiclass rail passenger revenue management (MPMC-RPRM) problem in this chapter. The major difference between our problem

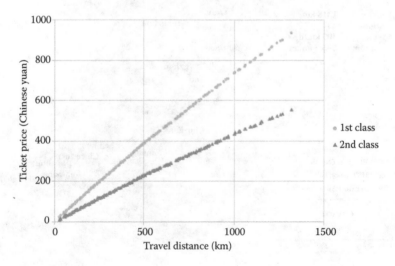

FIGURE 9.2
Ticket price of Beijing–Shanghai high-speed railway.

and the others in the RPRM literature is that we assume a portion of the unsatisfied demand from the previous period contributes to the demand of the next period. Similarly, we assume a portion of the unsatisfied demand from one seat class is transferred to the adjacent seat classes. By considering the spilled demands between periods and seat classes, we provide a more realistic description of passenger booking behavior. We then try to maximize the BSHR revenue by (1) optimizing the long-term daily train schedule and (2) allocating the seats to different legs of each train.

9.3 Problem Definition and Basic Formulation

In this section, we first provide a mathematical definition of MPMC-RPRM. Then we present a basic model (BM) to the problem. We also provide a trip-based model (TBM), which can be viewed as a Dantzig–Wolfe reformulation of BM and explain the reasons for the weak LP relaxation of TBM and BM.

9.3.1 Problem Definition

In this chapter, we consider a rail alone a set of stations S, starting from s_O and ending at s_D. Define $S' = S \backslash \{s_O, s_D\}$. There is a set of L legs, each leg $l \in L$ connecting two adjacent stations. We offer J possible journeys, each journey j is defined by a pair of origin–destination stations, and contains one or more connecting legs. Define J_l as the set of journeys containing leg l. We have a set of discretized time T, and a train can only depart from s_O at a time $t \in T$. For each train, we have a set of C seat classes, and the number of seats available for class $c \in C$ is n_c. The ticket price of a c-class seat for journey j is denoted as r_{cj}. In this research, we also make the following two assumptions.

All the trains depart from s_O and arrive at s_D. No trains can originate a trip from a station other than s_O, and no trains can terminate the trip at a station other than s_D. For simplicity, we use the term "trip" to represent a train traveling from s_O to s_D in the reminder of the chapter. For example, trip t is used to represent the train depart from s_O at time t.

The extra time for a train to make a stop at station $s \in S'$ is negligible, and the travel times of all the trains are the same regardless of the number of stops during the trip. Currently, the stopover time at any station $s \in S'$ is less than 3 minutes in BSHR. The average travel time is 331 minutes and the standard deviation is 19 minutes for all 66 trips. Therefore, it can be seen that assumption 2 is basically realistic. The underlying implication of assumption 2 is that if train A departs earlier than train B, train A will arrive at all the intermediate stations earlier than train B. This assumption is true for 56 out of all 66 trips for the BSHR, and for the remaining 10 trips, the train surpasses the previous train and arrives at s_D before the previous train for less than 30 minutes.

Because of the preceding two assumptions, we can define d_{cjt} as the demand of the c-class seats for journey j in the trip t. It is important to note that d_{cjt} is the demand accumulated from time $t - 1$ to t, assuming the demand before $t - 1$ has already been fully satisfied. In other words, we divide the demand of any journey j over the complete planning horizon into $|T|$ partitions, and in each partition, the demand is denoted by d_{cjt}. For example, assume we have a railway containing three stations A, B, and C, and a single class c, as shown in Figure 9.3. We have three possible trips whose departing times are $t_1 = 9{:}00$ a.m., $t_2 = 9{:}05$ a.m., and $t_3 = 9{:}10$ a.m., respectively. Let j_1 be the journey between station A and station C, and j_2 be the journey between station B and station C. Assuming the travel time of j_1 is 3 hours, and the travel time of j_2 is 1 hour. In Table 9.1, we show the demands of different journeys in different trips. Here, $d_{cj_1t_1} = 100$ represents that the demand of journey j_1 in trip 1 is 100 at 7:00 a.m.; $d_{cj_2t_1} = 150$ means that the demand of journey j_2 in trip 1 is 150 at 9:00 a.m. It is noted that the demand of j_2 is indexed by $t_1 = 7{:}00$ a.m. instead of 9:00 a.m. because the train that departing from station A at 7:00 a.m. will capture this demand. Also notice that $d_{cj_1t_2}$ is only 10 because it contains only the demand accumulated from 7:01 a.m. to 7:05 a.m. Similarly, $d_{cj_1t_2}$ is the demand accumulated from 7:06 a.m. to 7:10 a.m.

Then we define $\alpha_{ct(t+1)}$ as the spilled demand conversion rate from time t to time $t + 1$ for class c, when d_{cjt} cannot be fully satisfied. Now assuming we only

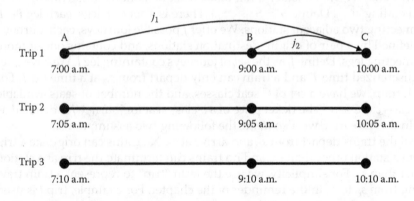

FIGURE 9.3
An example containing three trips between three cities.

TABLE 9.1

Demand of Three Trips at Different Times

t	d_{cj_1t}	d_{cj_2t}
$t_1 = 7{:}00$ a.m.	100	150
$t_2 = 7{:}05$ a.m.	10	5
$t_3 = 7{:}10$ a.m.	10	5

operate two trips at t_1 and t_3 in the example shown in Figure 9.3, the capacity of the train is $n_c = 80$, and $\alpha_{d_1 t_2} = \alpha_{d_2 t_3} = 1$, then the real demand of j_1 in the trip departing at t_3 is equal to $(\max(d_{d_1 t_1} - n_c, 0) \times \alpha_{d_1 t_2} + d_{d_1 t_2}) \times \alpha_{d_2 t_3} + d_{d_1 t_3} = (20 + 10) \times 1 + 10 = 40$. Similarly, we define $\beta_{cc't}$ as the spilled demand conversion rate from class c to c', when the demand d_{cjt} cannot be fully satisfied. One shall notice that $\alpha_{ct(t+1)} + \sum_{c' \in C \setminus \{c\}} \beta_{cc't} \leq 1$. $\alpha_{ct(t+1)}$ can be viewed as the demand conversion rate for the cost-sensitive passengers, and $\beta_{cc't}$ can be viewed as the demand conversion rate for the time-sensitive passengers.

For a trip departing at t, there is a fixed train operating cost, e_t. The objective of the proposed MPMC-RPRM problem is to decide which trips should be operated and the passengers allocated at each trip, so that the daily operational profit is maximized. Because the spilled demands between periods and seat classes are recaptured, we cannot decompose the problem into multiple single-period single-class RPRM problems directly. Therefore, we have to formulate the problem over the complete planning horizon.

9.3.2 Basic Model

To facilitate our discission, we first define the following variables.

- u_t: the binary variable such that $u_t = 1$ if the trip t is selected in the result, and 0 otherwise
- x_{cjt}: the number of the c-class seats allocated for c-class demand for journey j in the trip t
- $v_{c'cjt}$: the binary variable such that $v_{c'cjt} = 1$ if c-class seats are allocated for the spilled demand from class c' for journey j in the trip t, and 0 otherwise
- $y_{c'cjt}$: the number of c-class seats allocated for the spilled demand from class c' for journey j in the trip t
- z_{cjt}: the accumulated c-class demand for journey j in the trip t. It is worth mentioning that $z_{cjt} = d_{cjt}$ if there is no spilled demand from the previous time $t-1$; otherwise $z_{cjt} > d_{cjt}$.
- δ_{cjt}: the spilled demand of class c for journey j in trip t

Given the above variables, the BM for the MPMC-RPRM problem is formally defined as follows:

$$\max \sum_{t \in T} \sum_{c \in C} \sum_{j \in J} r_{cj} x_{cjt} + \sum_{t \in T} \sum_{j \in J} \sum_{c \in C} \sum_{c' \in C \setminus \{c\}} r_{cj} y_{c'cjt} - \sum_{t \in T} e_t u_t \qquad (9.1)$$

$$s.t. \sum_{j \in J_l} x_{cjt} + \sum_{j \in J_l} \sum_{c' \in C \setminus \{c\}} y_{c'cjt} \leq n_c u_t \quad \forall c \in C, \forall l \in L, \forall t \in T, \qquad (9.2)$$

$$z_{cjt} = d_{cjt} + \alpha_{c(t-1)t} \delta_{cj(t-1)} \quad \forall c \in C, \forall j \in J, \forall t \in T, \qquad (9.3)$$

$$\delta_{cjt} = z_{cjt} - x_{cjt} - \sum_{c' \in C \setminus \{c\}} y_{cc'jt} \quad \forall c \in C, \forall j \in J, \forall t \in T, \tag{9.4}$$

$$y_{c'cjt} \le \beta_{c'ct}(z_{c'jt} - x_{c'jt}) + M_{c'cjt}(1 - v_{c'cjt}) \quad \forall c \in C, \forall c' \in C \setminus \{c\}, \forall j \in J, \forall t \in T, \tag{9.5}$$

$$y_{c'cjt} \le M_{c'cjt}v_{c'cjt} \quad \forall c \in C, \forall c' \in C \setminus \{c\}, \forall j \in J, \forall t \in T, \tag{9.6}$$

$$x_{cjt}, y_{c'cjt}, z_{cjt}, \delta_{cjt} \ge 0, \text{ integers} \quad \forall c \in C, \forall c' \in C \setminus \{c\}, \forall j \in J, \forall t \in T, \tag{9.7}$$

$$u_t, v_{c'cjt} \in \{0, 1\} \quad \forall c \in C, \forall c' \in C \setminus \{c\}, \forall j \in J, \forall t \in T. \tag{9.8}$$

The objective function in Equation 9.1 is to maximize the overall profit, which is equal to the total revenue from tickets minus the total operational cost. The constraints in Equation 9.2 ensure that the total passengers traveling through any leg l of class c is less than the capacity of the train. The constraints in Equation 9.3 ensure that the accumulated demand of c-class for journey j at time t contains two parts, the original demand for period t and the spilled demand from the previous period $t - 1$. The constraints in Equation 9.4 compute the spilled demand of class c. Together with the non-negative constraints of variable δ_{cjt}, we also ensure that the number of the c-class seats for journey j of time t is less than or equal to the accumulated demand z_{cjt} at time t. We assume that $\delta_{cjt} = 0$ when $t = 0$ to avoid the confusion at the boundary condition. Equations 9.3 and 9.4 can be viewed as demand balance constraints, which is illustrated in Figure 9.4.

For each node z_{cjt}, we have two incoming flows, d_{cjt} and $\alpha_{c(t-1)jt}\delta_{cj(t-1)}$. Similarly, we have three outgoing flows for node z_{cjt}: they are x_{cjt}, $y_{cc'jt}$, and δ_{cjt}. The constraints in Equations 9.5 and 9.6 ensure that the converted demand from

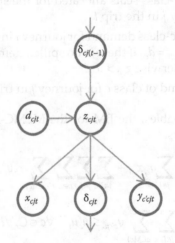

FIGURE 9.4
Demand flow balance between variables.

other classes to c-class is correctly computed, that is, $y_{c'cjt} = \beta_{c'ct} \max\{z_{c'cjt} - x_{c'jt}, 0\}$. Specifically, if $v_{c'cjt} = 0$, $y_{c'cjt}$ has to be 0; otherwise $y_{c'cjt} = \beta_{c'ct}(z_{c'jt} - x_{c'jt})$. Here, the smallest value of $M_{c'cjt}$ can be computed as follows.

$$M_{c'cjt} = \min\left\{n_c, \beta_{c'ct} \max\left\{0, \sum_{t'=0}^{t} \alpha_{c't't} d_{c'jt'} - n_{c'}\right\}\right\} \quad \forall c \in C, \forall c' \in C\backslash\{c\}, \forall j \in J, \forall t \in T.$$

$$(9.9)$$

Here $\alpha_{c't't}$ is the demand transaction rate from time t' to time t for class c', and $\sum_{t'=0}^{t} \alpha_{c't't} d_{c'jt'}$ computes the total accumulated demand from time 0 to time t for class c' of journey j. Therefore, $\max\left\{0, \sum_{t'=0}^{t} \alpha_{c't't} d_{c'jt'} - n_{c'}\right\}$ is the maximum possible number of unsatisfied demand at time t if trip t is selected in the solution. Therefore, $M_{c'cjt}$ is the maximum possible number of demand that can be transferred from class c' to c for journey j at time t. The constraints in Equations 9.7 and 9.8 are the integer and binary constraints for variables.

From our preliminary study, we find it is impractical to solve this model directly within a reasonable time for large test cases because there are a large number of binary/integer variables and big-M constraints. Therefore, we relax the integer constraints for (x, y, z, δ). We denote this relaxation as R(BM). The space complexity of this model is large, approximately $O(|C|^2|S|^2|T|)$ binary variables and $O(|C|^2|S|^2|T|)$ constraints.

9.3.3 Plan-Based Model and Analysis of the LP Relaxation

As we can see from BM in Equations 9.1 through 9.8, if we do not have constraints in Equation 9.4, the BM can be decomposed into T independent single-period multiclass problems. That is, the only linkage between T single-period multiclass problems is through the spilled demands. This motivate us to reformulate MPMC-RPRM problem in such a way that each variable in the new formulation represent a complete solution to a single-period multiclass RPRM (SPMC-RPRM) problem. In the reformulation, these variables are linked by the constraints that balance the demands of adjacent periods. We want to maximize the profit by selecting a single-period solution from each period.

Formally, define P_t as the set of all the possible solutions for a SPMC-RMRP on t. To avoid confusion, we use term "plan" to represent a single-period solution. Each plan $p \in P_t$, it contains detailed information about the accumulative demand z_{cjt}, seats allocated for the basic demand, and the spilled demand between classes x_{cjt}, and the spilled demand for the next period, δ_{cjt}. However, unlike in the BM where these values are the decision variables, these values are constants for any given p in the reformulation. To facilitate our discussion, we also define the following notations for each of the plan $p \in P_t$. Define the binary parameter u_p such that $u_p = 1$ if the trip of plan p is scheduled, and 0 otherwise. Define the binary parameter u_{ps} if the trip of plan p stops at station s. Define the nonnegative parameter x_{cjp} as the c-class seats for journey j in plan p. Define the nonnegative parameter $y_{c'cjp}$ as the spilled demand from c'-class to c-class for journey j in plan p. Define z_{cjp} as the accumulated c-class demand for journey j in plan p. Define δ_{cjp} as the spilled c-class demand for journey j in plan p. We then define the binary variable θ_p such that $\theta_p = 1$ if plan k is selected in the solution and 0 otherwise. The profit of plan p can be calculated as follows:

$$r_p = \sum_{c \in C} \sum_{j \in J} r_{cj} x_{cjp} + \sum_{j \in J} \sum_{c \in C} \sum_{c' \in C \setminus \{c\}} r_{cj} y_{c'cjp} - e_t u_p \quad \forall p \in P_t, \forall t \in T. \quad (9.10)$$

The mathematical formulation of the plan-based model (PBM) is then given by

$$\max \sum_{t \in T} \sum_{p \in P_t} r_p \theta_p \quad (9.11)$$

$$s.t. \sum_{p \in P_t} \theta_p = 1 \quad \forall t \in T, \quad (9.12)$$

$$\sum_{p \in P_t} \alpha_{ct(t+1)} \delta_{cjp} \theta_p + d_{cjt} = \sum_{p \in P_{t+1}} z_{cjp} \theta_p \quad \forall c \in C, \forall j \in J, \forall t \in T, \quad (9.13)$$

$$\theta_p \in \{0,1\} \quad \forall p \in P_t, \forall t \in T. \quad (9.14)$$

The objective function (Equation 9.11) is to maximize the overall profit. The constraints in Equation 9.12 ensure that one chain is selected for each period t. The spilled demand constraints shown in Equation 9.13 ensure the accumulated demand is from three sources, the spilled demand of the previous period, the original demand of the current period, and the spilled demand from other classes of the current period. There are $O(|C||S|^2|T|)$ constraints in this model, which is much less than the number of constraints in BM. In fact, it is not hard to see that PBM is a Dantzig–Wolfe decomposition of BM, and like many other Dantzig–Wolfe reformulation, PBM contains a large number of plan variables.

Despite the fact that there are millions of possible plans for any real-life test cases, Dantzig–Wolfe decomposition usually provides better LP relaxation to the original IP model. However, in our problem the LP relaxation of PBM is very poor. We illustrate the situation in the following example.

TABLE 9.2

Comparison Between Optimal LP Relaxation and Optimal IP Solution of PBM

Schedule	Depart or Not	Seat Allocated	Revenue r_p	Optimal IP	Optimal LP
p_1	No	0	0	1	0.3
p_2	Yes	100	200	0	0.7
p_3	Yes	70	−100	0	0

Consider a problem with a single class, a single journey, and a single period. The demand is $d = 70$, the train capacity $n = 100$, the ticket price $r = 10$, and fixed cost $e = 800$. We have three possible plans p_1, p_2, and p_3, as shown in Table 9.2.

It is obvious that the optimal IP solution is $\theta_{p_1} = 1$ and the revenue is 0. However, the optimal LP relaxation is $\theta_{p_2} = 0.7$ and $\theta_{p_1} = 0.3$, and the revenue of the LP relaxation is 140. This is because the revenue of the LP relaxation $\left(\dfrac{70}{100} \times r_{p_2} \right)$ is strongly affected by demand d in the LP relaxation of PBM. It is easy to see that when $d \le e/r$, the difference between optimal LP relaxation and optimal IP is equal to $d \times (nr - e)/n$, and the IP–LP gap is always 1. When d is slightly less than er, the LP relaxation of PBM provides little (and probably misleading) information on the optimal IP solution. In fact, this is also the case for BM.

9.4 A Decomposition-Based Two-Stage Heuristic

As we can see from the previous section, BM and PBM could provide poor LP relaxation to MPMC-RPRM and branch-and-bound has to be used to obtain the optimal IP solution, which could take a very long computational time. Therefore in this section, we propose a two-stage decomposition-based heuristic, in which we decompose each journey into a set of consecutive legs.

9.4.1 Decomposition-Based Model

In the decomposition model, instead of considering all journey J, we decompose every j into a set of consecutive legs denoted by L_j. We also approximate the revenue r_j by a set of r_l, where $l \in L_j$. As we discussed in Section 9.2, the ticket price r_j is affected by the travel distance linearly. Although there are two discount levels when the travel distance is more than 500 kilometers and 1000 kilometers, the number of journeys that are longer than 500 kilometers and 1000 kilometers is much less than the number of journeys that are shorter than 500 kilometers. For example, in BSHR, there are a total of 238 journeys and 138 of them are less than 500 kilometers, and only 24 journeys are longer than 1000 kilometers. Therefore, using r_l instead of r_j is a reasonable approximation for MPMC-RPRM. To facilitate our discussion, we define the following additional notations.

Additional notations:

d_{lt}: the accumulated demand for leg l of trip t. Here, $d_{lt} = \sum_{j \in J_l} d_{jt}$

r_l: the ticket price for leg l

Additional variables:

x_{clt}: the number of the c-class seats allocated for c-class demand for leg l in the trip t

$v_{c'clt}$: the binary variable such that $v_{c'clt} = 1$ if c-class seats are allocated for the spilled demand from class c' for leg l in trip t, and 0 otherwise

$y_{c'clt}$: the number of c-class seats allocated for the spilled demand from class c' for leg l in the trip t

z_{clt}: the accumulated c-class demand for leg l in the trip t

δ_{clt}: the spilled demand of class c for leg l in trip t

Given this notation, the decomposition-based model (DM) for MPMC-RPRM can be formulated as follows:

$$\max \sum_{t \in T} \sum_{c \in C} \sum_{l \in L} r_{cl} x_{clt} + \sum_{t \in T} \sum_{j \in J} \sum_{c \in C} \sum_{c' \in C \setminus \{c\}} r_{cl} y_{c'clt} - \sum_{t \in T} e_t u_t \qquad (9.15)$$

$$\text{s.t.} \ x_{clt} + \sum_{c' \in C \setminus \{c\}} y_{c'clt} \le n_c u_t \quad \forall c \in C, \forall l \in L, \forall t \in T, \qquad (9.16)$$

$$z_{clt} = d_{clt} + \alpha_{c(t-1)t} \delta_{cl(t-1)} \quad \forall c \in C, \forall l \in L, \forall t \in T, \qquad (9.17)$$

$$\delta_{clt} = z_{clt} - x_{clt} - \sum_{c' \in C \setminus \{c\}} y_{cc'lt} \quad \forall c \in C, \forall l \in L, \forall t \in T, \qquad (9.18)$$

$$y_{c'clt} \le \beta_{c'ct}(z_{c'lt} - x_{c'lt}) + M_{c'clt}(1 - v_{c'clt}) \quad \forall c \in C, \forall c' \in C \setminus \{c\}, \forall l \in L, \forall t \in T, \qquad (9.19)$$

$$y_{c'clt} \le M_{c'clt} v_{c'clt} \quad \forall c \in C, \forall c' \in C \setminus \{c\}, \forall l \in L, \forall t \in T, \qquad (9.20)$$

$$x_{cjt}, y_{c'clt}, z_{clt}, \delta_{clt} \ge 0, \text{integers} \quad \forall c \in C, \forall c' \in C \setminus \{c\}, \forall l \in L, \forall t \in T, \qquad (9.21)$$

$$u_t, v_{c'clt} \in \{0,1\} \quad \forall c \in C, \forall c' \in C \setminus \{c\}, \forall l \in L, \forall t \in T. \qquad (9.22)$$

Similar to Equation 9.23, $M_{c'clt}$ can be computed as follows.

$$M_{c'd} = \min\left\{ n_c, \beta_{c'd} \max\left\{ 0, \sum_{t'=0}^{t} \alpha_{c'rt}d_{c'lt'} - n_{c'} \right\} \right\} \quad \forall c \in C, \forall c' \in C \setminus \{c\}, \forall l \in L, \forall t \in T.$$

(9.23)

The space complexity of DM is much less than that of BM, approximately $O(|C|^2|S||T|)$ binary variables and $O(|C|^2|S||T|)$ constraints, which greatly reduces the difficulty of the model.

9.4.2 Two-Stage Heuristic

The results of DM provide the set of selected trips u_t; however, it cannot provide detailed information on how many seats should be allocated to each journey j. To resolve this issue, we fix the variable u_t in BM based on the results from DM, and then resolve the restricted BM. From our preliminary result, the restricted BM can be solved in very short time, and we can get the desired information on how many seats should be allocated to each journey j.

9.5 Model Extensions

In this section, we discuss two possible extensions of the proposed model. In particular, we first extend the model to handle multiple fleets. Then we also extend the model to incorporate the social welfare consideration.

9.5.1 Multiple Fleets

The traditional fleet assignment problem for aircraft or vehicles is to assign a variety of aircraft or vehicle fleets to individual flights or jobs based on passenger demands, revenues, operating costs, and so forth, so that the total profit is maximized (Hane et al., 1995; Sherali et al., 2006; Dumas et al., 2009; Liang and Chaovalitwongse, 2013). On the other hand, in the traditional rail industry, maximizing the operational revenue is often achieved by separating and recombining locomotives and cars at various locations in the rail network based on the demands (Cordeau et al., 2001; Lingaya et al., 2002; Cacchiani et al., 2010). However, a high-speed train is built more like an aircraft rather than a traditional train, because a car of a high-speed train cannot be easily attached or detached in the operation. Different fleets of high-speed trains have different capacities and operation costs. Therefore, instead of considering a single fleet, it is natural to extend our proposed model to handle multiple fleets, so that the total operational revenue is maximized.

The BM can be extended to handle the multiple fleets. This is done by introducing a new set of available fleets, I, and increasing the dimensionality of parameters and variables with fleet index. For each train in fleet $i \in I$, we have a set of C^i seat classes, and the number of seats available for class c^i is n_c^i. For a train of fleet i departing at t, there is a fixed train operating cost e_t^i. Define u_t^i as the binary variable such that $u_t^i = 1$ if the train of fleet i departing at time t is selected in the result schedule, and 0 otherwise. Define $v_{c'cjt}^i$ as the binary variable such that $v_{c'cjt}^i = 1$ if c-class seats are allocated for the spilled demand from class c' for journey j in the train of fleet i departing at t, and 0 otherwise. Define x_{cjt}^i as the number of the c-class seats allocated for journey j in the train of fleet i departing on t. Define $y_{c'cjt}^i$ as the number of c-class seats allocated for the spilled demand from class c' for journey j in the train of fleet i departing at t. The multifleet basic model (MFBM) for the MPMC-RPRM problem is given by

$$\max \sum_{i \in I} \sum_{t \in T} \sum_{c \in C} \sum_{j \in J} r_{cj} x_{cjt}^i + \sum_{i \in I} \sum_{t \in T} \sum_{j \in J} \sum_{c \in C} \sum_{c' \in C \setminus \{c\}} r_{cj} y_{c'cjt}^i - \sum_{i \in I} \sum_{t \in T} e_i u_t^i \quad (9.24)$$

$$s.t. \sum_{i \in I} u_t^i \le 1 \quad \forall t \in T, \quad (9.25)$$

$$\sum_{j \in J_l} x_{cjt}^i + \sum_{j \in J_l} \sum_{c' \in C \setminus \{c\}} y_{c'cjt}^i \le n_c^i u_t^i \quad \forall i \in I, \forall c \in C, \forall l \in L, \forall t \in T, \quad (9.26)$$

$$\delta_{cjt} = z_{cjt} - \sum_{i \in I} x_{cjt}^i - \sum_{i \in I} \sum_{c \in C \setminus \{c\}} y_{c'cjt}^i \quad \forall c \in C, \forall j \in J, \forall t \in T, \quad (9.27)$$

$$z_{cjt} \le d_{cjt} + \alpha_{c(t-1)t} \delta_{cj(t-1)} \quad \forall c \in C, \forall j \in J, \forall t \in T, \quad (9.28)$$

$$y_{c'cjt}^i \le \beta_{c'ct} \left(z_{c'jt} - x_{c'jt}^i \right) + M(1 - v_{c'cjt}^i) \quad \forall i \in I, \forall c \in C, \forall c' \in C \setminus \{c\}, \forall j \in J, \forall t \in T, \quad (9.29)$$

$$y_{c'cjt}^i \le M v_{c'cjt}^i \quad \forall i \in I, \forall c \in C, \forall c' \in C \setminus \{c\}, \forall j \in J, \forall t \in T, \quad (9.30)$$

$$x_{cjt}^i, y_{c'cjt}^i, z_{cjt}, \delta_{cjt} \ge 0 \quad \forall i \in I, \forall c \in C, \forall c' \in C \setminus \{c\}, \forall j \in J, \forall t \in T, \quad (9.31)$$

$$u_t^i, v_{c'cjt}^i \in \{0, 1\} \quad \forall i \in I, \forall c \in C, \forall c' \in C \setminus \{c\}, \forall j \in J, \forall t \in T. \quad (9.32)$$

The MFBM extends the BM presented in Equations 9.1 through 9.8 by introducing a new dimensionality of fleet index i to the load-related decision variables x and y. Also, the MFBM has a set of additional assignment constraints in

Equation 9.25 to ensure that only one fleet can be selected at any given period. The above model can be easily adapted to DM and the proposed two-stage heuristic.

9.5.2 Incorporating the Social Welfare

To achieve not only the profitability of BSHR, but also the economic and social impact on the population along the BSHR, it is important to ensure a certain number of stopovers at every city alone the BSHR. Economists and transportation experts cite CRH as one reason for China's continued economic growth. Although it might be unprofitable to make any stopover at a city from a financial point of view, such stopovers are considered crucial for the economic and social development of the regions. There are two ways to incorporate such the social welfare in the proposed model. We could directly maximize the social welfare as a part of objective function. However, it is very hard to provide an accurate estimation on the value of social welfare, and the objective function could become highly nonlinear and the models become very difficult to solve computationally. Therefore, our approach is to impose a set of hard constraints, so that the minimum number of stopovers at every city has to be more than a predetermined lower bound. This can be easily achieve in BM as follows.

$$u_t \geq w_{st} \quad \forall s \in S', \forall t \in T, \tag{9.33}$$

$$\sum_{c \in C} \sum_{j \in J_s} x_{cjt} + \sum_{c \in C} \sum_{c' \in C \setminus \{c\}} \sum_{j \in J_s} y_{c'cjt} \leq M w_{st} \quad \forall s \in S', \forall t \in T, \tag{9.34}$$

$$\sum_{t \in T} w_{st} \geq K_s \quad \forall s \in S', \tag{9.35}$$

$$w_{st} \in \{0,1\} \quad \forall s \in S'. \tag{9.36}$$

Here, K_s is the minimum number of stopovers at station s. The constraints in Equation 9.33 ensure that trip t can stop at station s only if trip t is selected in the solution. The constraints in Equation 9.34 are the logic constraints that ensure if station s is the departing or arrival station of any on board demand, w_{st} must be equal to 1. The constraints in Equation 9.35 ensure that the total number of stopovers has to be greater or equal to K_s. The constraints in Equation 9.36 are the binary variable constraints.

9.6 Computational Results

In this section, we report empirical results of the proposed models using a Dell Precision T7600 workstation with two INTEL Xeon E5-2643 CPUs of

3.3 Ghz, and 64 GB of memory on a 64-bit Windows 7® platform. Computational times reported in this section were obtained from the laptop's internal timing calculations. All the mathematical modeling and algorithms were implemented in MS Visual C++ 2010. All LP and MIP problems were solved using a CPLEX callable library version 12.5 with a default setting.

9.6.1 Test Case Generation

The realistic test instances used to benchmark the proposed models are created by utilizing the historical ticket selling records for BSHR by China Railway Corporation from March to May 2014.

In particular, we construct two test sets with 44 and 132 periods. Because the daily planning horizon is 11 hours, from 7:00 a.m. to 6:00 p.m., the duration between adjacent possible departures for the two test sets are 15 minutes and 5 minutes respectively. For each test sets, we create two test instances with 11 and 22 legs, respectively. The test instance with 11 legs contains 10 largest cities along the BSHR, and the test instance with 22 legs passing through all 21 cities along the BSHR. It can be seen that the largest test cases are larger than the current daily BSHR in terms of the number of legs and the number of possible periods. We consider only two seat classes, the first class and second class, in our test cases. The spilled demand transfer ratio from first class to second class is 0.15, and from second class to first class it is 0.05.

The daily demand of the test instances are obtained by computing the average daily from the historical data. We then use a mixture of two independent normal distributions to model a morning peak centered on 9:00 a.m. and an evening peak on 6:00 p.m. to model for daily demand (as shown in Figure 9.5). The 9:00 a.m. peak has a weight of 60% and the 6:00 p.m. peak a weight of 40%.

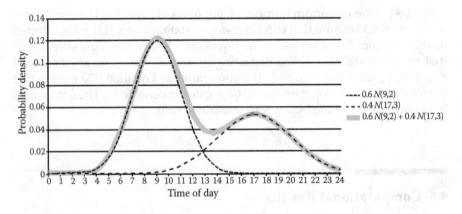

FIGURE 9.5
Demand over time.

We assume that the demand loss follows an exponential distribution. Specifically, in our model we assume that the unsatisfied demand decays by half every 2 hours, which is equivalent to losing 2.85% of the demand every 5 minutes. Therefore, the spill demand transfer ratios are 97.15% for 5-minute-interval test cases and 91.69% for 15-minute-interval test cases. We assume that the $e_t = 400k$ Chinese yuan for all $t \in T$. For each test instances, we also create three scenarios, in which the demands are set to 0.8, 1.0, and 1.2 times the average historical demand. Therefore, we have all together 12 test scenarios, and each scenario is named by pattern $T-A-B-C$, in which A is the number of legs, B is the number of periods, and C is the demand ratio factor.

9.6.2 Computational Results on the Basic Model

In this section, we report the performance of the basic model. The computational time is set to be 2 hours. In Table 9.3, we record the number of variables, binary variables, and constraints of BM. We also provide best IP solutions, best LP bound, IP–LP gap, and the computational time of BM. Here, the IP–LP gap is computed as IP − LPIP × 100%. All the computational times are recorded in seconds.

As we can see from Table 9.3, BM cannot obtain an optimal solution for any test scenario within 2 hours of computational time, possibly because there are a large number of binary variables and constraints in BM. The average IP–LP gap is 6.62%, ranging from 1% to more than 14%. The IP–LP gaps increase significantly when the number of periods increases. The demand ratio also affects the IP–LP gap greatly. The IP–LP gap increases significantly when the demand gets lower.

9.6.3 Computational Results on Two-Stage Heuristic

Table 9.4 presents the performance characteristics of the two-stage heuristic. The performance characteristics are (1) the problem size of DM (columns, rows, and binaries); (2) the computational results of DM (IP solution, best LP bound, IP–LP gap, computational time); and (3) the computational results of restricted BM (second stage of the two-stage heuristic).

As we can see from Table 9.4, the number of columns, rows, and binary variables is much less than that of BM. We can solve all 44-trip test scenarios in less than 6 minutes in the first stage of the heuristic using DM. However, the computational time increases drastically with the number of trips in the problem. We can only get a near-optimal solution for all 132-trip test scenarios. The average IP–LP gap for 132-trip scenarios is 3.33%. In the second stage of the heurisitc, all the restricted BM can be solved optimally within 1 minute. The objective value between DM in the first stage of the heuristic and restricted BM in the second stage of the heuristic is 14.61%. There could be two reasons for this. First, a large number of demands are traveling from

TABLE 9.3

Performance of BM

Test Scenario	Number of Columns	Number of Binaries	Number of Rows	IP Solution	LP Solution	IP–LP Gap (%)	Solution Time	Number of Trips
T-11-44-0.8	29,524	6292	25,080	3357	3554	5.87	7200	14
T-11-44-1.0	29,524	6292	25,080	4344	4491	3.40	7200	18
T-11-44-1.2	29,524	6292	25,080	5306	5406	1.89	7200	22
T-11-132-0.8	88,572	18,876	75,240	3293	3649	10.82	7200	14
T-11-132-1.0	88,572	18,876	75,240	4265	4603	7.92	7200	18
T-11-132-1.2	88,572	18,876	75,240	5097	5629	10.43	7200	21
T-22-44-0.8	112,288	23,232	92,840	4063	4197	3.29	7200	17
T-22-44-1.0	112,288	23,232	92,840	5189	5292	1.98	7200	22
T-22-44-1.2	112,288	23,232	92,840	6338	6412	1.17	7200	27
T-22-132-0.8	336,864	69,696	278,520	3951	4521	14.44	7200	17
T-22-132-1.0	336,864	69,696	278,520	5104	5629	10.28	7200	21
T-22-132-1.2	336,864	69,696	278,520	6222	6713	7.90	7200	27

TABLE 9.4

Performance of the Two-Stage Heuristic

Test Scenario	First-Stage DM							Second Stage		
	Number of Columns	Number of Binaries	Number of Rows	IP Solution	LP Solution	IP–LP Gap (%)	Solution Time	IP Solution	Solution Time	Number of Trips
T-11-44-0.8	4884	1012	4840	3966	3966	0.00	76	3378	8	14
T-11-44-1.0	4884	1012	4840	5096	5096	0.00	98	4325	9	18
T-11-44-1.2	4884	1012	4840	6208	6208	0.00	86	5285	8	22
T-11-132-0.8	14,652	3036	14,520	3943	4089	3.72	7200	3354	12	14
T-11-132-1.0	14,652	3036	14,520	5062	5200	2.73	7200	4295	11	18
T-11-132-1.2	14,652	3036	14,520	6167	6325	2.57	7200	5250	11	22
T-22-44-0.8	9724	1980	9680	4753	4753	0.00	327	4075	22	17
T-22-44-1.0	9724	1980	9680	6069	6069	0.00	217	5187	28	22
T-22-44-1.2	9724	1980	9680	7384	7384	0.00	90	6321	25	27
T-22-132-0.8	29,172	5940	29,040	4794	5036	5.05	7200	4120	29	17
T-22-132-1.0	29,172	5940	29,040	6165	6363	3.22	7200	5285	29	22
T-22-132-1.2	29,172	5940	29,040	7442	7645	2.74	7200	6401	38	27

Beijing to Shanghai, and the real ticket price (553 Chinese yuan) and the approximated price by legs (606 Chinese yuan) are different. Another reason is that because all journeys are decomposed into legs, the solution of DM has more seats × kilometers sold than that of BM. In BM, if any leg of a journey violates the seat capacity constraints, the ticket for the entire journey cannot be sold. However, in DM we can still sell the unaffected legs even if some legs of the journey violate the seat capacity constraints.

9.6.4 Comparison of the Results

In this section, we compare the solutions of BM, two-stage heuristic, and a genetic algorithm (GA) in Table 9.5. The GA has been used successfully in many large-scale combinatorial optimization problems. Therefore, it is natural for us to compare the proposed two-stage heuristic with a general heuristic such as GA. The detailed implementation of GA is given in the Appendix. The gaps for the two-stage heuristic and GA reported in Table 9.5 are computed based on the BM solution; for example, gap for the two-stage heuristic is computed as

$$\frac{BM\_Solution - Heuristic\_Solution}{BM\ Solution}.$$

As we can see from Table 9.5, the two-stage heuristic provides the best solutions on average. Particularly, the two-stage heuristic obtains the best solutions for 8 out of 12 test scenarios. For these 8 test scenarios, the average improvement from the BM solution is more than 2%. For the remaining 4 test scenarios, the average gap between the BM and two-stage heuristic is less than 0.3%. When the test scenarios are large, the two-stage heuristic performs much better than BM.

TABLE 9.5

Comparison of BM, Two-Stage Heuristic, and GA

Test Scenario	BM		Two-Stage Heuristic			GA		
	Value	Time	Value	Gap (%)	Time	Value	Gap (%)	Time
T-11-44-0.8	3357	7200	**3378**	0.62	84	3163	−5.76	7200
T-11-44-1.0	**4344**	7200	4325	−0.44	107	4140	−4.69	7200
T-11-44-1.2	**5306**	7200	5285	−0.38	94	4933	−7.03	7200
T-11-132-0.8	3293	7200	**3354**	1.87	7200	2842	−13.69	7200
T-11-132-1.0	4265	7200	**4295**	0.70	7200	3855	−9.61	7200
T-11-132-1.2	5097	7200	**5250**	3.01	7200	4532	−11.08	7200
T-22-44-0.8	4063	7200	**4075**	0.31	349	3447	−15.15	7200
T-22-44-1.0	**5189**	7200	5187	0.00	245	4681	−9.79	7200
T-22-44-1.2	**6338**	7200	6321	−0.26	115	5534	−12.68	7200
T-22-132-0.8	3951	7200	**4121**	4.30	7200	3401	−13.93	7200
T-22-132-1.0	5104	7200	**5285**	3.56	7200	4344	−14.89	7200
T-22-132-1.2	6222	7200	**6401**	2.88	7200	5468	−12.12	7200

Note: The best solutions are in bold.

Even for small test scenarios, when the total demand is low, the two-stage heuristic performs significantly better than BM. GA performs worst on all the test scenarios. Also, it is noted that the computational time of the two-stage heuristic for 44-trip scenarios is much less than that of BM and GA.

9.7 Conclusions

In this chapter, we studied a MPMC-RPRM problem. Different from other research work in the literature, the problem we studied recaptures the spilled demands between different time intervals and different classes. We present a basic mathematical model for the problem. Because there are a large number of binary variables and big-M constraints, it is hard to solve the real-life large-scale problem using the proposed model in a reasonable time. Therefore, we proposed a two-stage decomposition-based heuristic to solve the problem. We also demonstrate that the proposed models can be extended to model some real-life operational consideration. The computational results show that proposed two-stage decomposition-based algorithm outperforms the basic model and GA significantly.

In this chapter, we make the two assumptions. First, all the trains have to depart from s_O and arrive at s_D. In the real operations, it is possible for a train to start or end at a station other than s_O or s_D. By doing so, more demand might be captured and more revenue might be generated. Thus, extending the proposed model to handle this situation becomes an interesting and important future research direction. Second, we assume that the extra time for a train to make a stop at station $s \in S \backslash \{s_O, s_D\}$ is negligible. Therefore, it could be impractical to implement such a schedule because the later train might surpass the previous train in real operations. Therefore, integrating the detailed time-tabling decisions in the proposed model undoubtedly becomes an important future research direction. Finally, in our model, we have only considered the constant demand. In-depth work to handle the stochastic demand most likely will lead to better profits, and remains to be tested in our future study.

Appendix

GA has been used to solve many nondeterministic polynomial time (NP-hard) combinatorial optimization problems. Therefore, we propose a GA to solve MPMC-RPRM. The basic idea of our GA is that in each individual of the population, a subset of trips are fixed to 1 ($u_t = 1$), and then we solve the restricted

BM for the best possible objective value. The objective value is used as the fitness of the individual. Intuitively, the proposed GA tries to search the best set of trips (u_t) in the solution. Here, we use the list of binary value $u_t, \forall t \in T$ as the chromosome of individuals. Therefore, the size of the chromosome is $|T|$.

The GA starts with the generation of an initial population. The value of the initial population is set to 200 in our computation. From our preliminary study, the computation time increases with the number of selected trips $\sum_{t \in T} u_t$. Therefore, we compute an upper bound on the number of trips that can be selected as $\sum_{t \in T} \sum_{l \in L} \sum_{j \in J_l} d_{cjt} |L| \times n_c$. In each generation, we always select the top 80 individuals as the parents for the next generation, and these top 80 individuals are also carried to the next generation.

In the crossover procedure, we try to avoid a situation in which two trips that are close to each other are selected. Intuitively, if two trips are close to each other, the later trip usually cannot capture sufficient demands. Once the crossover operator has been applied and the offspring population has replaced the parent population, the mutation operator is applied to the offspring population. Specifically, for each u_t in the chromosome, a t is randomly chosen with probability $pm = 0.05$, and the value of u_t is changed to $1 - u_t$. In our computation, GA terminates until 2 hours computational time is reached.

Acknowledgments

This research work was supported by the National Science Foundation of China (Grant 71422003, Grant 71201003, and Grant 71371140).

References

Armstrong, A. and J. Meissner. Railway revenue management: Overview and models. Technical Reports Working Paper 035, Lancaster University, 2010.

Aydin, N., S. I. Birbil, J. B. G. Frenk, and N. Noyan. Single-leg airline revenue management with overbooking. *Transportation Science*, 47: 560–583, 2013.

Barnhart, C., A. Farahat, and M. Lohatepanont. Airline fleet assignment with enhanced revenue modeling. *Operations Research*, 51(1): 231–244, 2009.

Bertsimas, D. and S. de Boer. Simulation-based booking limits for airline revenue management. *Operations Research*, 53(1): 90–106, 2005.

Bharill, R. and N. Rangaraj. Revenue managemnet in railway operations: A study of the Rajdhani express, Indian railways. *Transporation Research Part A*, 42: 1195–1207, 2008.

Bradsher, K. *Speedy Trains Transform China*. Available at http://www.nytimes .com/2013/09/24/business/global/high-speed-train-system-is-huge-success -for-china.html, 2013.

Cacchiani, V., A. Caprara, and P. Toth. Solving a real-world train-unit assignment problem. *Mathematical Programming*, 124(1–2): 207–231, 2010.

China.com. *Beijing-Shanghai High-Speed Railway: Set the Example of High-Speed Railway in China (in Chinese)*. Available at http://finance.china.com.cn/roll /20140107/2102813.shtml, 2014.

China Railway Corporation. *Statistical Information on National Railway*. Available at http://www.china-railway.com.cn/gkl/tjxx/, 2014.

Ciancimino, A., G. Inzerillo, S. Lucidi, and L. Palagi. A mathematical programming approach for the solution of the rail way yield management problem. *Transportation Science*, 33(2): 168–181, 1999.

Cirillo, P. and P. Hetrakul. Passenger demand model for railway revenue management. Technical Report MAUTC-2009-01, Univeristy of Maryland, 2011.

Cooper, W. L. and D. Gupta. Stochastic comparisons in airline revenue management. *Manufacturing & Service Operations Management*, 8(3): 221–234, 2006.

Cooper, W. L. and T. H. de Mello. Some decomposition methods for revenue management. *Transportation Science*, 41(3): 332–353, 2007.

Cordeau, J.-F., G. Desaulniers, N. Lingaya, F. Soumis, and J. Desrosiers. Simultaneous locomotive and car assignment at VIA Rail Canada. *Transportation Research Part B: Methodological*, 35(8): 767–787, 2001.

Crevier, B., J.-F. Cordeau, and G. Savard. Integrated operations planning and revenue management for rail freight transporation. *Transportation Research Part B: Methodological*, 46(1): 100–110, 2012.

Dumas, J., F. Aithnard, and F. Soumis. Improving the objective function of the fleet assignment problem. *Transportation Research Part B: Methodological*, 43(4): 466–475, 2009.

Dutta, G. and P. Ghosh. A passenger revenue management system (rms) for a national railway in an emerging Asian economy. *Journal of Revenue and Pricing Management*, 11: 487–499, 2012.

Feigenbaum, B. High-speed rail in Europe and Asia: Lessons for the United States. Policy Study 418, Reason Foundation, 2013.

Hane, C. A., C. Barnhart, E. L. Johnson, R. E. Marsten, G. L. Nemhauser, and G. Sigismondi. The fleet assignment problem: Solving a large-scale integer program. *Mathematical Programming*, 70(1–3): 211–232, 1995.

Hetrakul, P. and C. Cirillo. Accommodating taste heterogeneity in railway passenger choice models based on Internet booking data. *The Journal of Choice Modelling*, 6(1): 1–16, 2013.

Lan, Y., M. O. Ball, and I. Z. Karaesmen. Regret in overbooking and fare-class allocation for single leg. *Manufacturing & Service Operations Management*, 13(2): 194–208, 2011.

Liang, Z. and W. A. Chaovalitwongse. A network-based model for the integrated weekly aircraft maintenance routing and fleet assignment problem. *Transportation Science*, 47(4): 493–507, 2013.

Lingaya, N., J.-F. Cordeau, G. Desaulniers, J. Desrosiers, and F. Soumis. Operational car assignment at VIA Rail Canada. *Transportation Research Part B: Methodological*, 36(9): 755–778, 2002.

McGill, J. I. and G. J. Van Ryzin. Revenue management: Research overview and prospects. *Transportation Science*, 33(2): 233–256, 1999.

Sherali, H., E. Bish, and X. Zhu. Airline fleet assignment concepts, models and algorithms. *European Journal of Operational Research*, 172(1): 1–30, 2006.

Talluri, K. T. and G. J. Van Ryzin. *The Theory and Practice of Revenue Management*. New York: Springer Science+Business Media, 2004.

Wikipedia. *Beijing-Shanghai High-Speed Railway*. Available at http://en.wikipedia.org/wiki/Beijing-Shanghai_High-Speed_Railway, 2014.

You, P.-S. An efficient computational approach for railway booking problems. *European Journal of Operational Research*, 185(2): 811–824, 2008.

Zhang, A., X. Fu, and H. Yang. Revenue sharing with multiple airlines and airports. *Transportation Research Part B: Methodological*, 44(8): 944–959, 2010.

Index

Note: Page numbers ending in "f" refer to figures. Page numbers ending in "t" refer to tables.